Cleaning Up the Mess

Cleaning Up the Mess

Implementation Strategies in Superfund

Thomas W. Church and Robert T. Nakamura

The Brookings Institution
Washington, D.C.

RECEIVED

About Brookings

The Brookings Institution is a private nonprofit organization devoted to research, education, and publication on important issues of domestic and foreign policy. Its principal purpose is to bring knowledge to bear on current and emerging policy problems. The Institution was founded on December 8, 1927, to merge the activities of the Institute for Government Research, founded in 1916, the Institute of Economics, founded in 1922, and the Robert Brookings Graduate School of Economics, founded in 1924.

The Institution maintains a position of neutrality on issues of public policy. Interpretations or conclusions in Brookings publications should be understood to be solely those of the authors.

Copyright 1993 ©

THE BROOKINGS INSTITUTION

1775 Massachusetts Avenue, N.W., Washington, D.C. 20036

Library of Congress Cataloging-in-Publication data

Church, Thomas W.
   Cleaning up the mess : implementation strategies in
Superfund / Thomas W. Church and Robert T. Nakamura.
      p.   cm.
   Includes bibliographical references and index.
   ISBN 0-8157-1414-9 (cloth)—ISBN 0-8157-1413-0 (pbk.)
   1. Hazardous wastes—United States—Management—
Finance—Case studies. 2. Liability for hazardous
substances pollution damages—United States—Cases.
3. Hazardous waste sites—United States—Case studies.
4. Pollution—United States—Case studies. 5. Hazardous
wastes—Law and legislation—United States.
I. Nakamura, Robert T.   II. Title.
   HC110.P55C48   1993
   363.72'876'0973—dc20   92-39493
                                                        CIP

987654321

The paper used in this publication meets the minimum requirements of the American National Standard for Information Sciences—Permanence of paper for Printed Library
Materials, ANSI Z39.48-1984

To our mothers,
Alice and Emiko,
and to the memory of our fathers,
Thomas and Kenzo

Contents

Preface

THIS BOOK is about Superfund, America's unique statutory scheme to use the more draconian elements of tort law to compel private businesses and public entities to clean up hazardous waste sites. The book is complex, befitting a program that may have generated as many acronyms as the New Deal (note the glossary at the back of this book). But our objective in undertaking this research was simple: to learn about the strategies used by the United States Environmental Protection Agency (EPA) to get individuals, corporations, and units of state and local government to pay millions of dollars to clean up environmental messes that these parties had a role in creating. We also hoped to offer some conclusions about the relative performance of the various approaches to this task that are authorized in the statutes.

Our objectives determined the approach we took to the research: a series of case studies of Superfund cleanups. These excursions into the nitty-gritty of decisionmaking at actual hazardous waste sites make up the central chapters of this book and provide the empirical basis for our subsequent analysis. This methodology limits the scope of the advice we ultimately are able to offer policymakers, however: while our concluding chapter offers some speculative judgments of the overall design and structure of Superfund, our basic endeavor has been to describe how the statutory and organizational framework operates and to provide some advice on how it might function more effectively.

Superfund has moved beyond its initial (and, some might add, exceedingly lengthy) shakedown phase. The basic legal tools are in place and their legitimacy has been established; an infrastructure of contractors has been created and many of the technical and scientific questions concerning site remediation are settled; the costs of creating the requisite bureaucratic apparatus in the EPA have been paid; and there are existing models for guidance on everything from the broad question of how the potentially responsible parties (PRPs) at new sites should be approached, to narrower issues involving the "boilerplate" of consent decrees and the administrative mechanics for setting up various types of funding arrangements.

While Superfund is finally up and running in a programmatic sense, the policy environment remains in flux. Congress's hurried and unheralded four-year extension of the program in 1990 did not follow the extensive reexamination that many had expected. Rather, it seemed more a reflection of political exhaustion after the acrimony accompanying passage of the Clean Air Act of that same year, and a desire to postpone a serious evaluation of Superfund until energy levels in the environmental policy community could be restored. In the meantime, a major insurance company continues to wage a public relations campaign in American newspapers and magazines, arguing that Superfund's liability-based system should be replaced by a public-works-style cleanup program financed by revenues from a proposed tax on commercial liability insurance.[1] Numerous reports on Superfund have appeared, several highly critical of various aspects of the program.[2] Others ask if cleanup funds could not be spent in more effective ways to reduce environmental risks.[3]

Decisionmaking at some Superfund sites seems to have satisfied the relevant constituencies, and participants in those successes express a degree of satisfaction with what transpired. Other sites represent what can only be described as slowly healing wounds, where all involved found the process leading to remediation needlessly slow, painful, and costly. The number of sites where even preliminary agreement has been reached concerning the basic structure of the final cleanup is small indeed in comparison to the National Priorities List (NPL) of the country's more than twelve hundred most hazardous toxic waste sites. The most dangerous sites have been rendered at least temporarily benign through emergency removal actions. A few are on their way to final cleanup. But in terms of the more permanent remediation envisaged in the statutory framework, Superfund's greatest challenges lie in the future.[4]

Within this context, we begin our analysis of the Superfund program. Enough is established so that extrapolations from the recent past can be a useful guide to an evaluation of the effectiveness of the various tools in the EPA's Superfund kit. At the same time, sufficient dissatisfaction with the status quo exists so that the research can inform those contemplating future change and adjustment.

Our primary focus is the decisionmaking process that precedes actual site cleanup. We have concentrated our attention on the statutory devices used to bring about the cleanup, the formal and informal strategies employed by the government in dealing with PRPs, and the implications these devices and strategies have for the resolution of the legal, monetary, and scientific issues that affect site cleanup.[5]

Surprisingly, the implications of the EPA's decisions on how to employ its arsenal of cleanup tools are underresearched. There is no shortage of reports advancing one enforcement strategy or another. Some suggest that a tough, coercive approach is most likely to produce sure and speedy settlements with PRPs.[6] Others suggest that a more conciliatory stance will produce the best results.[7] But these recommendations typically are based on a priori assumptions about individual and organizational behavior, the dynamics of negotiation and settlement, and the incentives of the participants. Few suggestions are grounded in empirical data regarding how these various strategies work in practice. Indeed, there is a paucity of published studies that systematically describe decisionmaking at one or more Superfund sites in enough detail to allow analysis and evaluation.[8] This research is an effort to fill some of the information gaps concerning operation of the Superfund program and to use this information to suggest how program implementation might be improved. At bottom, we hope to shed some light on that central question, what works?

We are indebted to a number of individuals and organizations for making this book possible. Clean Sites made a generous grant to the Research Foundation of the State University of New York, which supported both the field work and the subsequent analysis on which this book is based. The Center for Policy Research at SUNY/Albany and the university's Rockefeller Institute of Government provided administrative assistance. Our colleague, Phillip Cooper of the University of Kansas, made a particularly important contribution. Phil participated in the gestation of the study; he also did the field work for chapter 5 and wrote its first draft.

We would also like to acknowledge the consistently useful advice provided by Thomas Grumbly and Nancy Newkirk of Clean Sites; the help and support of Nancy Davidson and Theresa Walker at the Brookings Institution; the assistance of three of our graduate students at SUNY—Chris McMahon, Bong Lee, and William Wilkerson; and the comments and advice we received from Gene Bardach, Mark Landy, Kate Probst, Kent Stormer, and a host of others. Vincent Ercolano edited the manuscript, Susan Woollen prepared it for typesetting, and Max Franke prepared the index.

We owe perhaps our greatest debt to the dozens of members of the Superfund community, in and out of government, inside and outside the Beltway, who agreed to talk to us about this extraordinary program, and about their experiences in the cases we describe in chapters 3, 4, and 5. We cannot thank them by name because we promised them anonymity. We hope, however, that they will recognize some of their insights and perspectives in the pages that follow.

Though we have received much assistance, the views presented in this book are our responsibility and should not be ascribed to any of the persons or organizations acknowledged above or to the trustees, officers, or staff of the Brookings Institution.

Thomas W. Church
Robert T. Nakamura

Part One
Program Design

One

Introduction

SUPERFUND ALMOST certainly is the most expensive environmental program in the history of the United States, or of any other nation. With the 1990 reauthorization and its accompanying tax on chemical manufacturers, the cumulative authorized commitment of federal tax dollars to Superfund stands at $15.2 billion.[1] State and local funds, along with mandated contributions by the industrial parties made liable for cleanup costs by the statutes, will probably total many times the federal expenditure.[2] One estimate originally believed to be excessively pessimistic set the total cost of the Superfund program at $100 billion.[3] More recently, others have suggested that the "plausible upper bound" of total societal expenditures necessary to bring about an environmentally rigorous cleanup of all the nation's hazardous waste is a mind-boggling $1.7 trillion.[4]

The Superfund program is noteworthy not only for its substantial cost, but also for its unique design. It provides the United States Environmental Protection Agency (EPA) with an extraordinary variety of policy tools for accomplishing the mission of site remediation. The agency can order the private parties and corporations legally responsible for the toxic substances at a site—the potentially responsible parties or PRPs—to clean it up; this order can be either judicial or administrative, and can carry large fines and treble damages for failure to comply. In addition to potent sticks of regulatory authority, the agency is authorized to offer conciliatory carrots to bring about "voluntary," privately financed cleanup: mixed-funding arrangements whereby the government picks up part of the cost; limited releases of

3

liability for future cleanup expenses; and help in allocating the contribution of the various PRPs at multiparty sites. Alternatively, EPA can operate in a public works mode, initiating a federally directed and funded cleanup paid out of the Superfund. It then can bring legal action to force the PRPs to compensate the government for the administrative, engineering, and construction costs associated with the cleanup.

In subsequent sections of this introductory chapter we outline the basic structure of the Superfund program and our methodology for analyzing it, and provide a road map for the chapters that follow.

Statutory Framework

The Superfund program's enabling legislation consists of two statutes: the Comprehensive Environmental Response, Compensation, and Liability Act of 1980 (CERCLA), and the Superfund Amendments and Reauthorization Act of 1986 (SARA).[5] As the original statute's title suggests, the act provides for a federal *response* to clean up inactive, and often abandoned, hazardous waste sites. This response can be financed by a revolving fund (the Superfund), financed in part by a tax on chemicals and in part by general tax revenues. The statutory scheme also provides *liability* provisions by which past and present owners and operators of waste sites, and the transporters and generators of any hazardous substances found therein, are held responsible for cleaning up the toxic sites they helped create. Further, if federal funds are expended in the cleanup endeavor, the act establishes legal mechanisms by which the responsible parties can be compelled to *compensate* the government for its expenditures.

The Superfund statutes complement the Resource Conservation and Recovery Act of 1976 (RCRA), a more conventional piece of legislation that addresses the transportation and disposal of toxic substances. RCRA's future-oriented approach to toxic waste is similar in conception to the regulatory schemes established under the Clean Air Act and Clean Water Act. Under RCRA, the EPA has authority to define particular wastes as "potentially hazardous to human health or the environment" and to promulgate and enforce rules on how those materials should be transported and disposed of. CERCLA, however, was explicitly designed to remedy the *present* effects of *past* behavior—by defining the parties responsible for cleaning up preexisting (and typically inactive) toxic waste sites, by providing mechanisms by which the EPA can compel such private action, and by authorizing the government to conduct and fund the cleanups itself, should PRP intransigence, a lack of viable PRPs, or emergency conditions make such actions necessary.

The authorizing legislation establishes a complex set of options and procedures the EPA can use to effect environmental remediation of hazardous waste sites. CERCLA is based on a legislative determination of private-party liability for cleanup, and it is clear that Congress envisaged a system in which most of the cleanup work would be paid for, if not conducted by, responsible parties. The EPA also has authority to conduct investigations, request court orders that require responsible parties to clean up sites, and issue unilateral administrative orders to accomplish the same purpose. Noncompliance with these orders is punishable by substantial daily fines and treble damages.

Recognizing the potential for delay inherent in complex litigation involving substantial sums, Congress also created a public works component of the statute. A revolving fund was created to allow the federal government to conduct emergency cleanups at sites where an imminent danger to the public health or the environment exists. This fund also can be used to pay for more permanent remediation of non-emergency sites when the responsible parties are recalcitrant or where no financially viable parties are available. If viable responsible parties are present at such sites, the act provides the EPA with legal tools to recover the costs of remediation.

In facing the problem of how to clean up a toxic waste site, then, EPA officials have a number of formal decision points throughout a complex, statutorily mandated process.[6] If the problem can be defined as an emergency, the agency can proceed expeditiously with a *removal action* that addresses the immediate dangers. This largely unheralded aspect of Superfund has been responsible for the relatively speedy elimination of most of the worst dangers posed by hazardous waste in America. Typically, removals consist of what their name implies: drums, the hazardous contents of vats, pools, or other storage facilities, or (in extreme circumstances) surface soil are physically removed from a site and taken to a RCRA-authorized hazardous waste facility. Other protective actions also can be labeled emergency removals. These range from building fences around a site to creating an alternate water supply for nearby residents.

Superfund statutes place two important restrictions on the EPA's use of its removal authority. The statutes require that before a removal can be undertaken, a site must be found to pose an "imminent and substantial" danger to human health or the environment. They also place a statutory cap of $2 million on expenditures from the fund for any single removal action.[7] If viable PRPs are present at a site, the EPA can compel them to do the work (through issuance of unilateral administrative orders), or to compensate the government for expenses incurred in fund-financed removals.

At nonemergency sites, or at sites where an imminent danger has already been addressed through previous removal actions, the remedial mechanism is employed. Remedial actions are invariably complex and time-consuming. The most serious hazardous waste sites in the United States are placed on the National Priorities List through a process that assigns numerical values to various indicators of danger and potential for harm to health and the environment. These include size of site; levels and types of toxicity; categories of "media" (air, water, soil) that have been contaminated; and proximity to population centers, aquifers, and surface water. Although a site's ultimate ranking on the NPL is formally based on its score on a hazard ranking scale, the subjectivity of many indicators allows political variables to influence the final determination.

Once a site is listed, all remedial actions must proceed through a uniform, statutorily mandated process.[8] After the preliminary assessment that precedes listing on the NPL is complete, a two-part scientific and engineering study is conducted. The first part assesses the geography, geology, and hydrology of the site and surrounding area, and maps the location and toxicity of the various contaminants found. This study is called the *remedial investigation*. It is almost invariably linked to the *feasibility study*, an engineering report that sets out the estimated costs and benefits of the various options for cleaning up the site. This combined *remedial investigation/feasibility study*, or RI/FS, can take several months to several years to complete, and typically is conducted by one of a handful of environmental consulting firms around the country that perform this specialized work. Before an

RI/FS begins, the EPA usually notifies all the PRPs it can locate of their potential liability, and offers them the chance to conduct the study themselves, under agency supervision.[9] The agency may also issue an administrative order under section 106 of the statute, which requires the PRPs to conduct the study under pain of fines and treble damages. (An instance of this approach—the case of the Laskin Poplar Oil Site—is discussed in chapter 3.)

The RI/FS provides the scientific and legal basis on which the EPA regional administrator selects the remedial option which appears to best satisfy the broad, arguably conflicting, statutory requirements for cleanup. It should (1) adequately protect human health and the environment; (2) satisfy all relevant federal, state, and local environmental standards—the *applicable or relevant and appropriate* standards, or ARARs; (3) to the extent possible, utilize "permanent" solutions (as opposed to placing impermeable caps on, or fences around, waste sites) and new technology; and (4) be cost-effective. After requisite hearings and a comments period, the regional administrator factors in local political conditions and the mood at agency headquarters, weighs what the cooperating PRPs are likely to accept,[10] and formalizes his or her choice of remedial option in the *record of decision*, or ROD.[11]

Once the record of decision is filed, final design and actual construction of the remedy takes place in what is termed *remedial design and remedial action*, or RD/RA. Final design and construction work may be undertaken by the PRPs, pursuant to a consent decree or other formalized agreement with the EPA. The PRPs may be mandated to perform this work, under a unilateral order issued under section 106 of the statute. Or, the EPA may perform the work itself, drawing from the Superfund and reserving the right to take legal action to obtain compensation for its costs from PRPs. When a site is officially considered to have been cleaned up, it is formally *delisted* and removed from the NPL.

As this outline suggests, the process of remedial decisionmaking and actual cleanup can be time-consuming and highly expensive. Once remedial action is underway, the actual cleanup—particularly when it involves attempts to clean contaminated soil on site through "pump and treat" technology—can require a decade or more to complete. The best available figures suggest that the average Superfund remedial case takes forty-three months from the time the EPA becomes aware of the site until NPL listing, twenty months from that stage until commencement of the RI/FS, thirty-eight months more until issuance of the ROD, eighteen months for remedial design, and twenty-five months for completion of the remedial action. The average total time consumed is a little over eight years.[12] The average cost of remediating a site, not including legal expenses and other transaction costs, currently exceeds $30 million.[13]

Competing Strategies for Superfund Implementation

The lengthy statutory menu of choices provided by Congress, and augmented by the courts, gives the EPA a uniquely varied range of implementation choices. But the agency was provided little guidance on how it might choose among those options, or how it should resolve the tensions inherent in an effort to obtain speedy remediation, at a high level of environmental purity, with minimal expense to the federal government. Before much headway can be made in sorting out the respective costs and benefits of these alternative approaches to achieving the goals of the Superfund program, one must have a better understanding of how the program works in practice.

Previous empirical research on Superfund has touched on implementation issues, but has tended to focus on a scientific and technical evaluation of the remedies applied to particular sites. Relatively little is known about the costs, benefits, and preconditions for success of the alternative approaches implied in the statute. Fortunately, Superfund implementation has varied over time and also across regions—though more because of politics at the headquarters level and the centrifugal tendencies of a decentralized bureaucracy than because of any systematic attempt to find out which approach works best in a particular situation. The program thus provides a kind of laboratory in which different approaches to implementation can be examined in operation, and then compared and assessed.

We took advantage of this quasi-experimental situation to conduct an investigation of decisionmaking processes at several selected Superfund sites. It was our hope that detailed, analytical case studies of the use of different approaches to implementation would advance understanding of how the program operates in the real world. More important, we hoped that the case studies would illuminate some of the implications of choosing one or another of the possible approaches authorized in the statutes. While an in-depth look at a few cases cannot resolve the complex policy issues affecting the Superfund program, we hope that such an examination can place those issues in a real-world context.

The research presented in this book began with a series of loosely structured interviews in Washington, Philadelphia, and New York with members of the national Superfund community—lawyers, administrators, engineers, researchers, and representatives of interest groups who are actively involved in one way or another with the Superfund program. These interviews helped us develop our research strategy and determine if our reading of the statutes and the literature resonated with their perception of the program. The picture of Superfund that emerged from our reading was modified and clarified in light of these preliminary forays into the field.

Our preliminary interviews suggested that the EPA regions have developed distinctive approaches to the statutory tools available to them. Some regions have a reputation in the Superfund community for being prosecutorial: they "issue section 106 orders at the drop of a hat"; they are described by PRPs as uncooperative, unaccommodating, and adversarial. Other regions are seen as more "reasonable," more "helpful," more "creative in trying to work out a settlement." Finally, at least one region has a reputation for being run on a public works basis. In this region, "getting on with cleanup" is the order of the day; it is said that "they don't let lawyer bickering stand in the way of cleaning up sites."

This regional variation resonates with changes in Superfund implementation over time. The EPA leadership under successive presidential administrations has pressed for one or another of these three broad strategies for cleaning up waste sites. Anne Burford, EPA administrator during the first two years of the Reagan administration, used her legal discretion to shape a Superfund program based on conciliation with private industry. The Burford era was characterized by an emphasis on restrained use of federal cleanup money, voluntary agreements with responsible parties, and a corresponding federal willingness to settle for levels of cleanup deemed by some to be inadequate. The pejorative term for the agreements reached in this era, widely used by members of the environmental community, is "sweetheart deals," and allegations of political manipulation of Superfund cleanups were a major factor in Burford's forced resignation from the agency. The poor reputation of this period has complicated the position of advocates of a more conciliatory approach toward business.

Burford's successors, beginning with William Ruckelshaus, responded to these criticisms by stressing a more aggressive, litigation-based strategy, including increased referrals to the Department of Justice for enforcement actions and more reliance on unilateral administrative orders. Polluters were identified as "bad guys" and the EPA saw its primary role in terms of law enforcement.

The third strategy is typified by the efforts of Winston Porter, Superfund administrator in the late 1980s. Porter's background was in engineering, and his emphasis was on getting cleanup operations under way, leaving disputes over who would pay for subsequent resolution by lawyers and accountants. Superfund monies for both remedial and emergency actions were spent aggressively. The strategy of preserving the fund through PRP-led cleanups gave way to the desire for quick action through a public works approach to environmental remediation.[14]

These observations suggest that at least three approaches or strategies have been used to implement the Superfund program. The first emphasizes coercion. The law is used to get a defined population to do something it would not otherwise do—clean up sites—through application of techniques ranging from the threat of having to pay for a government-conducted cleanup, to administrative and court orders to perform specified site-remediation actions, with the accompanying threat of heavy fines and treble damages in the event of noncompliance. Because of its retrospective focus on the "bad acts" of responsible parties, we term this implementation strategy *prosecution*.

The prosecution approach to Superfund implementation is frequently advocated by environmental groups and members of Congress.[15] It is well-summarized in an unpublished report commissioned by the EPA in the late 1980s that urged the agency to adopt "an effective enforcement program," one in which "the government create[s] a reasonable expectation in those legally obligated that fulfillment of the legal obligations will be less costly than failure to fulfill the obligations."[16] The report continued by specifying how to attain this goal, in terms more reminiscent of the hyperbole of criminal justice than of regulation. (Emphasis is in original.)

In order to create this expectation, it is necessary that the government accomplish three essential tasks. It must: *detect* the individual members of the community subject to the law who are not fulfilling their obligations; *punish* those individuals; and *communicate* to the public and the obligated community the requirements of the law and the consequences of not fulfilling those requirements.[17]

The second implementation strategy is based on what has been termed a cooperationist model of government-business relations.[18] Despite its coercive elements, the legislation authorizing Superfund is explicit in promoting negotiation between the agency, the PRPs, and other concerned parties. These statutory provisions suggest that Congress envisioned an approach in which the PRPs are given credit for at least a modicum of good faith, site cleanup is perceived as an activity requiring cooperative business-agency relationships, and negotiated dispositions are viewed as preferable to litigation. We call this the *accommodation* strategy, a term that suggests the shared interests of government and the PRPs, and emphasizes the need for all sides to resolve differences in good-faith negotiations rather than through adversarial procedures.

Support for the accommodation strategy is found at several places in the statutory framework. There was, for example, explicit encouragement of alternative dispute resolution in SARA, Superfund's first reauthorization statute.[19] The EPA is authorized to provide PRPs with "negotiation moratoria" before commencement of the remedial investigation/feasibility study or the actual cleanup. Section 104 of the act authorizes private responsible parties to conduct both the RI/FS and actual remedial action, with EPA approval. The EPA can prepare a nonbinding allocation of responsibility (NBAR) as an aid in settlement discussions among PRPs. The Superfund statute also authorizes the government to accept settlements that embody a number of specific elements calculated to increase private incentives to settle: *de minimis* buyouts (whereby parties with small proportionate contributions to a waste site are enabled to pay a fixed sum and thereby obtain a complete release from future liability), mixed funding (an arrangement by which cleanup costs are shared by PRPs and the Superfund), releases from future liability through an EPA covenant not to sue, and through protection against lawsuits by other PRPs. Indeed, SARA requires the government to furnish private PRPs with a written statement of the reasons for rejection of any "substantial" settlement offer.

The EPA's national approach to Superfund has oscillated between prosecution and accommodation, depending on agency leadership and the external political environment.[20] The agency also has access to a policy tool that sidesteps the tension between these two approaches altogether. The statutes clearly authorize EPA to encourage, coerce, or compel private parties to clean up toxic waste sites. But Congress also gave the agency a huge pot of money and a mission: clean things up. The agency therefore has both the authority and the financial resources to clean up sites itself, without PRP participation. We term this alternative the *public works* approach to Superfund implementation. It finds support in the statutes, in the actions of various past administrators of the Superfund program, in the policies pursued by several of the semi-autonomous regions of EPA, and in the suggestions of some special-interest groups.[21] The public works approach, which underlies the emergency removal provisions of the statutes, embodies an engineering perspective on cleanups. It suggests that the goal of site cleanup is poorly served by a process in which decision costs can equal or even exceed actual cleanup expenses. Rather, the operant call is to "bring in the bulldozers," and let the lawyers and accountants fight over the bill later.

These alternative approaches to Superfund cleanups, while analytically distinct, often blur in practice. The government rarely pursues a "pure" prosecution, accommodation, or public works strategy at any particular site. Enforcement activities in the real world often involve mixed strategies, beginning with removal actions conducted and financed by the government (public works), followed by alternating combinations of tough talk and threats to PRPs (prosecution) and amiable encouragement of a negotiated settlement (accommodation). However, discussion of the costs and benefits of unmixed enforcement options has animated much of the public debate over the program, and over the proper stance to be taken by the EPA in the future. More important, our research suggests that frequently it is possible to identify relatively clear *orientations* toward one or another of these alternatives in government actions, both in specific cases and more generally in EPA regions. Thus, even though the categories of prosecution, accommodation, and public works may overlap somewhat, we believe that they represent meaningful alternative strategies which in fact guide the decisions of government personnel as they implement the Superfund program.

A Case Study Approach to Superfund

The central chapters of this book summarize the six case studies that resulted from our research, organized according to the EPA regions in which the sites are located. We discuss in this section our case-selection strategy and the reason for this organization by EPA region. Superfund cases are managed initially by semi-autonomous regional offices that are largely insulated from policy shifts in Washington. Headquarters is always involved in politically visible cases, but much of the Superfund program operates at the regional level without much supervision from the center. While there are limits to how far a region can proceed in settlement negotiations without headquarters approval, the critical early days of a case are very much in the control of the region.

While not all regions have a distinct persona, our informants across the system generally agreed that a prosecutorial, litigation-based strategy characterized Regions II (New York, New Jersey, and Puerto Rico) and V (the Great Lakes states); a settlement-oriented approach was emphasized in Region X (the Pacific Northwest), and especially in Region III (the Middle Atlantic states). Region IV, encompassing most of the states of the Deep South, reportedly took a strong public works approach to Superfund sites.

Our interviewees suggested a variety of causes for these regional differences: the personality and occupational background of regional administrators;[22] the size, complexity, and configuration of PRPs at typical Superfund sites; and the constraints placed on program implementation by state environmental policy, including the availability of state funds to pay for the statutorily mandated share of remediation expenses at so-called fund-lead sites in which government funds are used to finance cleanup activities.[23] The putative causes of this variation are of less relevance to our effort, however, than its existence.

If our interviewees had been less consistent in their characterizations of the general approaches of the EPA regions, we would have been less willing to select our case studies on a regional basis. Given the near-unanimity of their assessments, however, we decided to study two cases from each of three regions: III, IV, and V. These regions together contain almost 50 percent of the sites on the National Priorities List. They constitute three of the top four regions in the country in terms of numbers of NPL sites.[24] As an intermediate check on our characterizations, we asked regional officials how they viewed their own approach to Superfund implementation, and how they perceived the approaches of the other regions.[25] Again, we found a reassuring degree of consistency.

While we are reasonably confident in the general accuracy of our descriptions of regional Superfund personalities, our goal has not been to provide an up-to-date portrait of regional policy and practices. We went to Regions III, IV, and V because we wanted to find real-world examples of the accommodation, public works, and prosecution approaches to Superfund. Fortunately, we did in fact find Superfund cases that were broadly illustrative of these three strategies in the regions where we expected to find them. But we are not implying that *all* Superfund cases in each of these regions are conducted on the basis of the regional characterization. Furthermore, subsequent changes in personnel or policy at the national or regional level may have lessened the validity of these descriptions.

With two sites to be selected in each of three regions, we were left with the possibility of varying the cases along another dimension. Our preliminary interviews suggested that an important factor differentiating Superfund sites is the overall level of difficulty in achieving a site cleanup—in monetary, technical, legal, and political terms. Some sites reportedly are "hard"; some are "easy." While practitioners do not always agree on what distinguishes hard from easy sites, they do believe that these two categories provide a useful distinction that explains much of the variance in decisionmaking processes at different Superfund sites. We therefore decided to investigate one hard site and one easy site in each of the three regions we had chosen.

Some of the major factors our interviewees believed differentiate hard and easy sites include: the documentation available concerning the source, quantity, and character of hazardous material; the number, size, and cooperativeness of the PRPs; the estimated costs and technical complexity of the cleanup; and the cleanup's political visibility. While the hard-versus-easy distinction did not admit of precise definition, it proved less of a problem in actually differentiating sites. In a manner analogous to Justice Potter Stewart's approach to obscenity,[26] we had some difficulty defining the difference between hard and easy sites, but when we got into the field, we knew them when we saw them. Many of the factors contributing to site difficulty or easiness traveled together. Thus, large sites with many PRPs typically involved higher estimated cleanup costs, more technical problems, and higher political visibility. The converse can be said in broad terms regarding smaller sites with fewer PRPs.

The hard sites we selected were large dump sites with many PRPs, high estimated cleanup costs, and a variety of technical and engineering problems. The easy sites involved few PRPs, far lower estimated cleanup costs, and no special political, technical, or evidentiary difficulties. Because we were most interested in the full span of decisions leading up to the actual remediation, we selected only sites in which remedial design, though not necessarily remedial action, had been completed.

The cases were constructed through a combination of documentary research and interviews with participants. In each case, we sifted through the administrative record, including the RI/FS, ROD, and any other crucial milestones in the case. Where court proceedings were instituted, we also scanned the court files. This documentary research preceded a series of interviews conducted with the participants in decisionmaking at each site. Where possible, we interviewed the remedial project manager (RPM) in charge of the site, the relevant assistant regional counsel, and their supervisors in the regional office. We also interviewed lawyers representing the major PRPs and, in one case study, the federal district judge overseeing the case.

The interviews were conducted on a not-for-attribution basis: we promised subjects that nothing said in the interview would be attributed to them. Superfund cases seem never to be completely resolved; while we chose cases in which the basic decisions about remedy and financing had been made, most still had potential litigation of one sort or another pending when we conducted our interviews. We thus were fortunate to obtain the cooperation of the attorneys and administrative personnel involved in the cases. We doubt that this cooperation would have been forthcoming if our interviews had been on the record.[27]

In chapter 2 we describe the statutory tool kit provided to the EPA by the Superfund legislation, setting out in more detail the alternative strategies of prosecution, accommodation, and public works, and providing a discussion of the various evaluative standards by which program success can be assessed. Chapters 3, 4, and 5 consist of the case studies compiled in the course of our research. In the final two chapters we discuss the lessons to be derived from the case studies, and their broader implications.

Two

Superfund's Tool Kit

ENVIRONMENTAL POLLUTION poses a classic example of what economists term *market failure:* individuals pursuing self-interested objectives collectively produce unsought and undesirable outcomes for society as a whole. Public intervention frequently is used to reallocate the costs and benefits of such dysfunctional economic activity in socially desirable directions.

Toxic cleanup is a technically uncertain enterprise in which the costs are extraordinarily high, the benefits uncertain, and the source of financing subject to heated controversy. The very goals of the undertaking can be obscured in a maze of political questions masquerading as issues of scientific judgment, and by a multiplicity of conflicting objectives (such as speed, environmental effectiveness, and economic efficiency) encompassed by the seemingly unambiguous concern with "cleanup."

The Superfund program reflects these dilemmas. For policy analysts, the enabling legislation is interesting not simply because it addresses a peculiarly troublesome and complex environmental issue, but also because it gives the Environmental Protection Agency unusually broad discretion in defining both the means and the ends of the program. Superfund provides the EPA with a uniquely varied kit of alternative statutory tools that can be combined to produce quite different cleanup programs. Agency discretion in the use of its policy implements has, in fact, created substantial variation in implementation strategy across regions and administrative "eras." The statutory framework also establishes a set of goals and performance indicators that, because of their number and inherent tensions, virtually require the agency to craft its own definition of programmatic success in Superfund. This chapter establishes a framework for subsequent examination of Superfund's tool kit in operation: the policy tools, implementation strategies, and programmatic goals that underlie America's hazardous-waste cleanup effort.

Policy Tools

An emerging body of research on the strengths, weaknesses, and behavioral and contextual assumptions of the various implements of government action provides a general perspective on the issues of policy design underlying Superfund.[1] This work focuses on "the instruments common to different policies and on the conditions under which these instruments are most likely to produce their intended effects."[2] Although the typologies of policy implements put forward by various researchers differ at the margins, they all include variants of the same basic alternatives. This research suggests that government can:

—Regulate, or otherwise force compliance, through "command and control" techniques

—Create incentives for compliant behavior

—Exhort and persuade

—Tax and spend

Interestingly, Superfund's primary statutory implement is omitted from existing typologies of generic policy tools: modifying civil liability doctrine in order to alter the legal rights and duties of concerned parties. Superfund imposes on a defined set of responsible parties a legal duty to clean up waste dumps—a responsibility that can be enforced through civil action by the government in federal court.

As a policy tool, modification of civil liability is not unique to Superfund; it has long been used to achieve public purposes. For example, liability of atomic power plants for potentially catastrophic injuries to neighbors of the facilities was statutorily limited in the 1950s to encourage private investment; a similar grant of qualified civil immunity was later granted to the producers of several vaccines against childhood diseases.[3] No-fault insurance schemes, workers' compensation, and other forms of "tort reform" increasingly have been used to effectuate public purposes through modifications of private-party liability for civil wrongs. These examples illustrate what might be termed *negative* changes in civil liability, instances in which legislatures have limited the enforceable rights of private parties. On the positive side, statutes augmenting civil rights have included provisions that create new private causes of action enforceable in court, aimed at newly defined harms such as employment discrimination or racial harassment.[4] Superfund is unlike either of these more typical forms of policy-driven uses of civil liability doctrine, however; Superfund creates a right of government to compel privately financed cleanups, and a corresponding duty on a defined set of private parties to comply.

This liability-changing aspect of the Superfund program is sometimes confused with regulation. We were told in several interviews within the EPA, for example, that administrators view the authorizing cleanup statutes as broadly analogous to the Clean Air Act and Clean Water Act, or the Resource Conservation and Recovery Act (RCRA) regulating waste-disposal practices. There are superficial similarities: costs are imposed on a defined segment of the population, while the benefits are enjoyed by the larger public. The EPA must define key elements of the program through traditional administrative processes, and has a battery of enforcement tools analogous to those used in regulatory programs.

However, the Superfund program differs in key aspects from regulation. On one hand, regulatory activity is nearly always directed at influencing the present and future behavior of the targeted population. Superfund, on the other hand, assigns liability for past actions.[5] In addition, regulators usually are required to apply general rules to a population defined in terms of its present commercial activity. Because of its focus on abandoned sites, Superfund requires the EPA to create a list of potentially responsible parties through investigation of past acts that may have no direct or obvious relationship to current business operations. This factor creates incentives for potentially responsible parties at Superfund sites to "hide in the weeds," an option that is seldom available when regulation is applied to ongoing enterprises. Furthermore, the EPA must define technical standards and goals anew for each cleanup; each requires a unique determination of how clean is clean?, and the technologies to be employed to reach that objective. Administrative rule-making in regulation, however, involves the formulation of standards of more general applicability.

Superfund's liability scheme embraces a standard common to

much environmental regulation: the "polluter pays" principle.[6] Particularly in times of financial constraint, it seems appropriate that those who contribute to environmental ills should pay to have them corrected. However, three key issues confound this approach when it is applied to the cleanup of abandoned hazardous waste sites, and serve to differentiate Superfund's use of liability doctrine from most previous applications of the directive that the polluters should pay:

1. *The standard of liability or fault applied.* In claims for personal injury in most legal systems, a finding of fault or negligence on the part of the defendant is necessary before responsibility is incurred. However, a liability standard based on negligence complicates the assignment of legal responsibility for dump sites, since evidence of negligent dumping frequently is hard to come by. The alternative to a traditional fault-based standard—strict liability—eliminates this difficulty by assigning responsibility without the need to prove negligence, but it achieves this objective at the cost of potential unfairness to law-abiding and otherwise responsible waste disposers.

2. *The apportionment scheme used to allocate responsibility at sites to which several parties contributed hazardous waste.* In multiple-party dump sites, it is frequently impossible to determine whose waste contributed to which damage. Yet, most legal systems stress the necessity of finding a direct causal link between a particular defendant's actions and a specific injury. If any recovery is to take place at multiple-party hazardous waste sites, it usually must be based on a method of apportioning liability that sidesteps the frequently intractable problem of establishing this linkage. Superfund's solution to this problem—joint and several liability—allows full responsibility for harms caused by multiple parties to be assessed against any one defendant found to have contributed to those harms. Dilution of the causality requirement in tort law, however, is inevitably accompanied by potential unfairness to defendants held responsible for injuries they did not cause.

3. *The treatment of dumps created prior to enactment of the liability changes—the issue of retroactivity.* A determination must be made concerning the responsibility of private parties for sites created before liability was statutorily imposed. Since most cleanup programs are aimed at existing sites, a purely prospective liability doctrine would be effective only at those existing sites where defendants can be found liable under traditional legal standards. Yet retrospective changes in

liability raise obvious and particularly serious concerns of fairness and due process.

As finally interpreted by the federal courts, Superfund embodies the harshest responses to each of these quandaries: liability is strict, joint and several, and retroactive.

Superfund's redefinition of the legal obligations of hazardous waste generators, transporters, and dump owners is not the only statutory tool in EPA's Superfund kit. It is supplemented by a tax-and-spend, public works component for dealing with sites that constitute immediate public health or environmental emergencies. Government, acting on its own initiative, can initiate removal actions to address these emergency situations, and can also remediate less critical sites in the absence of viable responsible parties. If full assumption of remediation expenses by responsible parties is either unfeasible or otherwise inappropriate, monies from the Superfund trust fund can supplement private-party contributions through so-called mixed-funding arrangements.

These statutory tools are accompanied by a host of subsidiary instruments provided to the EPA: formal power to determine cleanup requirements at individual sites, to issue unilateral administrative orders to private parties, and to provide settlement incentives to encourage private parties to comply voluntarily with agency requests. In addition, when the EPA uses its public works authority to clean up sites on its own, it can subsequently file lawsuits against private responsible parties to recover its costs.

The tool kit provided to the EPA in the Superfund legislation contains a number of potent policy implements. Yet the agency was provided with no instruction manual describing how those tools should be used.

Implementation Strategies in Superfund

The policy tools defined in authorizing legislation are seldom self-implementing or even self-defining. Rather, they must be combined, interpreted, and applied within the constraints posed by the resources and the culture of the responsible government agency, and by other critical actors in that organization's environment.[7] Agencies, in other words, must determine what use to make of their kit of policy implements. In the process, they must also select the broad approach they will take to the individuals and organizations affected by agency behavior. This determination can be self-conscious or can just happen. But whether the process is implicit or explicit, government agencies necessarily adopt a *strategy* for using their available policy tools. With the exception of a burgeoning literature on strategies in the use of regulation, this process is frequently passed over in discussions of generic policy tools.[8]

Strategies are guides to action. They posit scenarios that begin with an actor's present situation and resources, then move through suppositions about what a series of indicated actions will yield at each step, and finally show how the connected steps will produce the desired result. In short, strategies guide actors by telling them how to invest their resources, orchestrate their actions, identify benchmarks to measure progress, and bring coherence and purpose to their cumulative activities.

Typically, implementation strategies are made up of several elements:

—A defining characteristic that can be easily described and understood;

—A theory about why a particular approach will produce the desired result;

—A notion of the series of tasks that must be performed, and how the agency should apportion its efforts;

—Guidance on how government actors should behave toward other participants, how they should present themselves by words and deeds to communicate how they may be expected to behave and how they expect others to behave.[9]

For implementation strategies to be effective, then, they must show program managers how to use existing statutory tools. This activity always takes place in an environment that poses a variety of constraints—economic, administrative, and political. Superfund's environment features many such constraints, but two are fundamental. First, the Superfund trust fund, despite its substantial size, is much too small to finance all the necessary cleanups. Second, the administrative personnel available to the EPA are limited in number, experience, and capacity; the agency simply cannot shoulder all the administrative and managerial tasks required to achieve expeditious cleanups of the required number of sites.

Any strategy for Superfund implementation must address these two resource-related constraints in the EPA's strategic environment. The unambiguous statutory solution to both, and thus a mandatory element in any implementation strategy in Superfund, is the liability scheme that requires individuals, businesses, and other government units to absorb many of the cleanup costs and administrative expenses at sites where they are the responsible parties. Any strategy for Superfund implementation ultimately must be grounded in the coercive tools that can force the statutorily defined responsible parties to assume cleanup costs. To borrow a colorful phrase from one of the attorneys we interviewed when planning this project, these tools constitute the EPA's "gorilla in the closet."[10] But different uses can be made of the gorilla. It can be kept in the background as an implicit threat, while program staff try to deal with responsible parties on a cooperative basis. Alternatively, the EPA can rattle the cage door frequently, even threaten to open it, in order to frighten the responsible parties into compliance. Finally, the agency might not resort to the gorilla at all until the cleanup is completed, at which point it could be used to ensure that the requisite bills are paid.

Our exploratory interviews within and outside the EPA suggested that each of these approaches to Superfund's liability provisions have been used at various times in the history of the program (See chapter 1 in this book). In addition, we uncovered a common belief among Superfund regulars that regional offices of the EPA have developed idiosyncratic Superfund "personalities" that mirror the categories of this rough typology.[11] We conceptualized these three alternative approaches to use of Superfund's liability doctrine as, respectively, accommodation, prosecution, and public works.

These three approaches address a common problem through reliance on the same set of policy tools. But they do so based on different assumptions and techniques. We now turn to a systematic discussion of these alternative strategies that organize our understanding of the EPA's implementation options in the Superfund program. The strategies are presented neither as ideal types—theoretical constructs divorced from observed reality—nor as concrete descriptions of actual implementation regimes observed in practice. Rather, we attempt to illuminate a middle ground between the prosecution, accommodation, and public works approaches as abstract ideas, and the case-specific material presented in subsequent chapters. Our goal is to clarify each of the implementation strategies and its implicit premises.

Prosecution

The essential characteristic of the prosecution approach is an emphasis on coercion and a reliance upon the legal power of the government to command compliance. The EPA's goal in the prosecution strategy is to compel "responsible" behavior on the part of potentially responsible parties who would not clean up sites unless the costs of noncompliance exceeded the costs of compliance.[12] The theory underlying this approach is that once the targeted PRPs recognize that they will inescapably be held individually responsible for carrying out all cleanup activities at a given site, they will find it in their interest to organize settlements to share costs, take responsibility for site management, and otherwise discharge their legal obligations.

The central tenent of the prosecution approach is that the selected PRPs can and must be made fully responsible for all cleanup costs. This is both the government's negotiating position and what frequently appears to be a moral imperative. In this approach, the government plays an uncompromising role analogous to that assumed by prosecutors dealing with especially blameworthy criminal defendants. Strict, joint and several liability in this context dictates that any deviation from assumption by the PRPs of all the costs and risks associated with cleanups constitutes an unwarranted, if not improper, concession to parties who have no legitimate right to escape the full implications of their legal position.

The prosecutorial perspective on PRP liability has been supported consistently by policymakers. Senator Robert Stafford, for example, made the case for a strong prosecutorial approach to Superfund cleanups when the Superfund Amendments and Reauthorization Act (SARA) was debated in Congress:

> Joint and several liability . . . shifts the burden of pursuing recalcitrants from the Government to other potentially responsible parties. The theory underlying Superfund's liability scheme was, and is, that the Government should obtain full costs of cleanup from those it targets for enforcement, and leave remaining costs to be

recovered in private contribution actions between settling and non-settling parties.[13]

Stafford subsequently stated, "Under a liability scheme as rigorous as that of Superfund, recoveries of amounts less than full costs are disfavored and are justified by only extraordinary circumstance."[14]

A similar point was made by the Environmental Law Institute, in an unpublished report to the EPA on how the agency might improve its Superfund enforcement program. This report argued that the EPA's concern with achieving settlements with PRPs is misguided, and leads to unnecessary expenditures:

> The existing Superfund program fails to achieve effective enforcement. The approach taken by the Agency is a "settlement-first" approach. It does not satisfy the important principles of effective enforcement. For example, instead of issuing administrative orders, a condition precedent to penalties and punitive damages, EPA concedes costs to lure and reward settlers. Because the Agency does not rely on a credible threat of punishment and increased costs to induce responsible parties to perform fully their statutory obligations, public money is spent which achieves little return in shaping responsible party behavior to achieve cleanup. . . . An enforcement program that is based upon aggressive use of powerful enforcement authorities approaches settlements differently. The government should use its resources to create the perception in members of the regulated community that it is more risky to refuse to fulfill their legal obligations than to fulfill them. To generate this perception, the Agency must prove its willingness to punish those who fail to fulfill their obligations. Through actions in other cases, around the country and over time, EPA can induce the regulated community to seek compliance and settlement on the government's terms. By developing and implementing an aggressive enforcement stance, the Agency is more likely to achieve settlements on terms favorable to the government at a minimum cost to it.[15]

The EPA itself points to the enforcement approach as an efficient way of producing cleanup dollars:

FY90 Superfund Enforcement produced major results for the investment . . . The Superfund Enforcement Budget . . . was $200 million. The value of enforcement activities was $1.6 billion. This represents an 8 to 1 return on the Enforcement investment. . . . The value of enforcement actions . . . equaled the size of the Superfund budget.[16]

In short, a prosecutorial approach provides a way of substituting coercive power for the agency's limited funds and administrative capacity.

ESSENTIAL REQUIREMENTS. To pursue a prosecutorial strategy, the EPA must assemble a "policy machine" to do several things.[17] First, it has to identify a target group of PRPs with the appropriate attributes. In cases involving a large number of PRPs, this involves selecting for prosecution a small and manageable group of experienced PRPs with the required monetary resources. Second, the agency must go after this select group with the threat or actual use of the coercive legal tools in its arsenal to compel compliance. This emphasis on the use of legal tools means that the agency must involve the Department of Justice, which represents it in court in most legal actions. Inevitably an element of control is lost to the Justice Department, but in the case of the prosecutorial strategy, the goals of the two organizations are largely congruent. Third, the EPA must conduct its other activities—scientific studies, removals, the various engineering studies, issuance of records of decision and administrative orders—in ways that can be defended in court.

The government's behavior in implementing a prosecution strategy is intended to give PRPs the impression of implacability. Because PRPs must be convinced that they have no choice but to comply, the government makes few concessions in negotiations. The government makes frequent use of administrative orders against PRPs to compel them (on pain of treble damages) to take desired cleanup actions. The government is unyielding in the exercise of its administrative prerogatives; for example, PRPs are given only a limited, formal opportunity to comment on and influence the specification of the work they are to be compelled to perform. Consultation with PRPs does not exceed the minimum requirements specified in the statutory framework.

In this adversarial relationship, the agency goal is to achieve capitulation by the targeted PRPs. It is only through their agreement to pay all past and future costs and their assumption of all future risk that targeted PRPs can start reducing their own share of costs (through orchestration of negotiated settlements with other nontargeted PRPs) and shaping the cost of remediation (through the design and conduct of the remediation, subject to EPA oversight).

PRESENTATIONAL STYLE. A definite presentational style is associated with the prosecution strategy. The style is suggested by its appellation—students of regulation label the relevant image "the regulator as policeman."[18] The key government actors in the prosecution approach are the lawyers—drawn from the EPA's regional counsel staffs and from the Justice Department—who orchestrate the legal maneuverings and conduct negotiations. Other participants—the remedial project managers and technical staff—may play important roles, but prosecutorial strategies put the lawyers in the lead.

The informal language of government lawyers is often tough and uncompromising, especially when referring to PRPs. They become "slam-dunk" PRPs or "deep pockets" (a term that we learned is offensive to industry representatives). The Justice Department's term for those proceeded against in Superfund cases is *responsible parties* because, as one Justice official told us, "We don't go after *potentially* responsible parties." One former high regional official commented that the practice of mixed funding is seen by some government lawyers as "immoral," since it lets PRPs stick taxpayers with costs that the PRPs should pay. Such views are an outgrowth of the professional training of lawyers, of the dictates of the arguments that they make, and of their interactions with the PRPs themselves.

Pending legal proceedings necessarily give an aura of defensiveness and secretiveness to the prosecution strategy. Lawyers and those they represent are reluctant to give out information voluntarily, particularly when the government must rely on sometimes questionable data from consultants, on internal records assembled in haste and amidst uncertainty, and on difficult negotiations with the states to produce official decisions which must in turn be defended in court. The adversary process, with its assumptions about self-seeking behavior, discourages unilateral candor and openness. Information is a resource useful in conflict and, as such, should be controlled and rationed for effective use.

While prosecutors present themselves as tough and effective adversaries, or as "policemen," they expect their opponents to behave as "amoral calculators."[19] PRPs will do what government expects only when the costs of compliance are lower than the costs of noncompliance. The goal of the prosecution strategy, then, is to convince PRPs of the applicability of this cost-benefit calculation to their own circumstances.

Accommodation

The accommodation approach to Superfund implementation stresses the mutual, rather than conflicting, interests of the EPA and the PRPs in site cleanup. Given the appropriate circumstances and behaviors, accommodation leaves both parties better off than if they had pursued more confrontational strategies. The strategy assumes that PRPs recognize their liability and the agency's potential to use its coercive tools. In this context, the theory that underlies accommodation suggests that most (or at least some) PRPs will want to discharge their legal obligations "voluntarily" and, given a correctly structured chance and a willing government partner, will cooperate in order to minimize their costs. A group of major PRPs made the case for accommodation in the following terms:

> The real issues of concern to EPA and industry are settlement, selection of an appropriate remedy, and equitable allocation of costs. By working with industry, EPA can help resolve these issues without expensive and protracted litigation so that the efforts of all parties can focus on quick and effective cleanup, and EPA will not have to use its limited resources on litigation.[20]

The difficulty, of course, is that PRPs and the EPA have both mutual and conflicting goals concerning Superfund. Neither side wants to pay more, or assume more risks, than it must. This is of particular concern to the EPA in a political climate in which appearing to be overly conciliatory toward "polluters" can have serious implications. The gamble taken by those who employ the accommodation strategy is that the divergent interests of the agency and the PRPs will be compromised in a mutually beneficial agreement that is achieved faster, with fewer transaction costs, than under a more adversarial regime.

ESSENTIAL REQUIREMENTS. The central task for the EPA when using the accommodation strategy is to create the circumstances under which PRPs will cooperate. Three related tasks follow from this objective:

—A common pool of information on which to base informed discussions must be generated.

—A relationship of trust must be built that prevents adversarial stances on both sides.

—Incentives must be provided to encourage PRPs to settle.

An accommodation strategy, then, requires the EPA to assemble its available discretionary powers and personnel resources to lay the foundations for performance of these tasks.

Program personnel have to provide a basis for understanding the technical problems presented at a site and to share that information with the PRPs. And since the hope is that negotiations will resolve the issue of who pays, the EPA and the PRPs should cooperatively develop an objective basis for apportioning shares of the costs, such as volumetric and toxicity data. Because of its investigatory powers, the EPA often can generate this information most efficiently; it is produced least efficiently through litigation. Insistence that strict, joint and several liability removes the government from any role in the allocation of cleanup costs among PRPs is likely to be unproductive in this context, since allocation of shares is typically the most difficult issue in Superfund negotiations.

If the goal is to build a cooperative relationship, then formal tasks become opportunities to build trust. That means using smaller occasions for cooperation (initial emergency removal actions, for example) as steps toward building cooperation for larger enterprises (such as conducting the requisite engineering studies). The government also should avoid changing its position dramatically, imposing new requirements, or taking other actions that would undermine the impression of trustworthiness and consistency. A minimum requirement for the development of trust, in other words, is for the agency to manifest a concern for some degree of procedural fairness in its dealings with PRPs.

Finally, the accommodation strategy requires willingness on the part of the government, as well as the PRPs, to make concessions. Even if the EPA holds all the cards in a legal sense, both sides must appear to be negotiating in good faith if the accommodation approach is to be successful. Concessions need not be substantial, depending on what the situation requires. The EPA may be asked to exercise discretion in ways that build trust, split at least a part of the difference on issues of importance to the PRPs, and otherwise indicate that its representatives are bargaining in good faith. The EPA can accomplish this objective by being flexible in negotiations over cleanup standards, by providing administrative help to the PRPs in reaching agreement among themselves, and by assuming some portion of the political and monetary risks involved in a cleanup.

PRESENTATIONAL STYLE. Like prosecution, accommodation has a characteristic presentational style. Whereas lawyers are the central actors in prosecution, in the accommodation approach the program staff is central.[21] Often program personnel have scientific backgrounds and tend to see the technical issues as primary; problem-solving becomes an exercise in which people representing different sides can cooperate to reach a mutually satisfactory solution.

Giving at least the impression of openness is an essential element of the presentational style of accommodation-oriented participants. While secretiveness is functional in an adversarial environment, it nurtures a mistrust that undermines accommodation. And while the prosecution strategy treats PRPs as "amoral calculators"—and includes a set of routines intended to reduce them to that status—accommodation represents an effort to nurture the "good-citizen" side of PRPs as businesses interested in doing the right thing in both their own and the public interest. The agency staff, instead of being policemen, become more like politicians.[22]

For accommodation to succeed, the EPA and the PRPs must have a mutual interest in settling quickly and getting on with cleanup work at the site. The ingredients for settlement—the compilation of technical data to determine what has to be done, and the gathering of volumetric and toxicity data to help PRPs apportion costs—should be within the capacity of the agency to deliver. The EPA has the legal authority to allow PRPs to participate in technical discussions; PRPs need to participate in these discussions cooperatively and constructively. Finally, both sides must be willing and able to share in some of the financial costs and future risks involved in obtaining a settlement.

Public Works

The essence of the public works approach is the segmentation of tasks: clean up first and recover costs second. It places a priority on action, focusing on those engineering tasks that are likely to be completed quickly, and postpones the more difficult process of negotiating with PRPs and allocating costs. Priority is assigned to those tasks that require the smallest transaction costs and delays, and for which the necessary means are readily available. Those tasks involving disputes over allocation of costs and the selection and implementation of long-term remedies are best delayed. This hierarchy of values promotes the use of removal actions, for which obtaining PRP agreement through complex negotiations is unnecessary, and for which the requisite funds, agency discretion, and technical capabilities are readily available.

The public works implementation strategy assumes that rapid action is desirable and that the EPA has the capacity to act quickly and effectively when the realm of activity is narrowly defined. A related assumption is that the EPA's limited resources are best deployed in doing what it can to improve the physical world first—by eliminating the most pressing threats to health and the environment—before worrying about either long-term remediation or cost recovery. At the least, the agency can reduce the scope of any remedial activity by aggressive use of removals. The principal advantage of this strategy is speed in getting cleanups started. However, this advantage is loaded at the front end of the process, with higher costs and more difficult activities concentrated toward the end.

We should distinguish, at this point, between what is normally considered a traditional public works approach to social problems and what we term the public works strategy for Superfund implementation. In traditional terms, public works programs apply government staff and funds directly to achieve a public policy objective. As such, public works is an important element on any list of the tools of public policy.

Public works programs entail few of the conceptual complexities of regulation or of attempts to change legal rights and duties. Rather, they involve the simple expedient of spending government money, under agency direction. Establishment of a traditional public works program has, in fact, been proposed as a possible reform of Superfund. Such a program would establish a larger, more broadly based cleanup fund that would pay for remediation at all sites, without reliance on private-party liability. This approach, combined with a prospectively imposed liability system for improper disposal of hazardous waste, is already used to clean up hazardous waste dumps in several European countries.[23]

Our use of the term *public works* relates not to the generic public policy tool, but to an implementation strategy that uses an existing set of statutory tools in a way that approximates a more traditional public works program, particularly at the front end of the process. Like the other implementation strategies we have outlined, public works promises to deliver on the existing goals of the Superfund program, which include minimization of taxpayer expense. This objective is attained through the subsequent use of liability doctrine to recover costs from responsible parties. We are interested in how this implementation strategy solves the same problem addressed by its alternatives: how to achieve cleanups paid for by PRPs.

ESSENTIAL REQUIREMENTS. The basic organizational, situational, and technical requirements of the public works approach are relatively simple. First, aggressive program and technical staff must be willing to interpret the EPA's emergency removal powers broadly. Removals require a finding of an "imminent and substantial threat to public health or the environment," and cannot exceed statutory limits in terms of total expenses. Those using the public works approach must interpret these constraints leniently. Second, there must be a ready capacity to implement inexpensive, low-tech solutions—primarily the transporting of hazardous materials to approved disposal sites.[24] Third, an adequate apparatus to conduct cost recovery activities (PRP searches, apportionment of costs, enforcement actions) is necessary, at least if the strategy is not to be a pure exercise in spending public funds.

In the public works model, lawyers take a secondary role to the program people and they operate primarily as bill collectors. Getting the PRPs to pay the bill is facilitated by the comparatively low cost of removals—their statutory restriction to $2 million may be flexible to a degree, but it is not infinitely so. Furthermore, since cleanup costs presumably have already been paid, little uncertainty remains about how much they will be. This situation contrasts sharply with that presented by prosecution and accommodation, both of which typically deal with PRP assumption of virtually unlimited, and highly uncertain, future costs.

PRESENTATIONAL STYLE. The presentational style associated with the public works approach is segmented, like the strategy itself. At the outset, it involves aggressive administrators at the top of the regional structure who are prepared to take the political risks associated with an aggressive use of Superfund monies to clean up sites. The ethos is one of "getting the stuff off the ground as soon as possible," a perspective that also may necessitate playing fast and loose with formal rules in order to achieve results on the ground. In the cost recovery phase, by contrast, the dominant role is played by lawyers who take either the neutral perspective of bill collectors or the tough posture of prosecutors, depending on circumstances and individual predilections.

A scenario for the success of the public works approach would look something like this: Once a site is selected, the EPA uses its emergency removal power to authorize quick action, usually by transporting ground-level contaminants to a nearby RCRA facility. As much of this work as possible is done under discretionary removal powers. These actions presumably reduce the problems at the site, increase knowledge of the overall dimensions of the remaining contamination, and provide the government with an incentive to get started on recovering fund outlays. For relatively small sites, these activities may sufficiently address conditions so that little else needs to be done. For bigger sites, the preliminary work provides some experience at the site, and subsequent negotiations with PRPs can be based on fuller information. In either case, cost recovery is treated as a mop-up operation following the initial cleanup action.

Implementation Strategies and Regulatory Strategies

The discussion of alternative implementation strategies in Superfund resonates with a larger debate in the policy community over regulation and the approach government regulators should take to private businesses. Despite its dissimilarities to regulatory schemes,

Superfund poses many of the same choices facing government officials charged with administrating traditional forms of regulation.

The major issue at most Superfund sites is whether public officials should negotiate and compromise with PRPs to reach mutually satisfactory resolutions of the issues that divide them, or take a more formal, adversarial, and explicitly coercive approach. The literature on regulation posits two similarly contrasting strategies. One is adversarial, procedure-bound, and coercive. The other involves more discretion for negotiated compromises, and relies upon cooperation. Ultimately, the choice of regulatory strategies is the choice of whether "to punish or to persuade."[25]

Recent studies of regulation have rekindled debate over the amount of discretion bureaucracies should have in enforcing the law. Critics of the more formalized, coercive approach suggest that it breeds "regulatory unreasonableness" and argue, with considerable force, that a harsh enforcement regime produces high transaction costs and an unproductive, acrimonious relationship between business and government.[26] This approach allegedly contributes to what Robert Kagan has called a culture of "adversarial legalism," in which the discourse shifts from policy concerns and programmatic objectives to arguments among lawyers over legalistic minutiae, and in which energies are diverted into processes stacked in favor of deadlock.[27]

Most criticism of regulatory unreasonableness has grown out of case studies of administrative agencies charged with delivering publicly valued goods—cleaner air, safer workplaces, more access for people with disabilities, less discrimination—by means of requiring changes in the behavior of various targeted groups, usually businesses. The costs of delivering these goods are borne by the regulated population, while the benefits are dispersed to a larger public. Since these studies often focus on the excesses of regulation, the case for more conciliatory approaches is made primarily by negative example.[28]

Interestingly, some older literature on agency discretion reached quite a different conclusion. For example, Theodore Lowi argued against giving agencies the ability to negotiate with regulated publics over regulatory requirements.[29] Grant McConnell contended that such discretion increased the role of private power at the expense of the public good.[30] This older literature is based on the evils of what was termed "agency capture," a situation in which government officials identify too closely with those they are supposed to be regulating. These concerns are directly analogous to the more recent arguments of public interest groups against administrative use of accomodationist strategies when dealing with the environment, consumer protection, and workplace safety. The preferred alternative is a tough enforcement approach that emphasizes coercion, frequent recourse to the courts, and formal, adversarial relationships between regulators and the regulated.

A frequent criticism of those who advocate a less confrontational regulatory system for the United States—such as that found in many European countries and in Japan—has been allegedly immutable cultural differences that make such "cooperationist" practices inappropriate in America.[31] In most discussions of regulatory approaches, the United States ranks at the adversarial endpoint of the continuum. But, as we have indicated, the strategies we describe are rooted in differences observed at different times and in different regions of the United States. In other words, while the United States may be a more adversarial society than many nations, there is still freedom for motion in one direction or the other.[32]

The foregoing indicates that a debate over government-business relations is taking place at several levels—among Superfund practitioners, proponents of "tort reform," and students of public policy. Some aspects of the debate are concerned with comparative cultural perspectives, other parts with theoretical advantages of one orientation over another, still another aspect with reactions to the excesses of one approach or another.

We are interested in a narrower concern. Rather than seeking to identify the most productive governmental approach to regulated businesses, our interest lies in identifying the best strategy for achieving Superfund's objectives. This effort requires some notion of programmatic success. Before moving to our case studies, therefore, we briefly discuss the various notions of success in the Superfund program.

Goals

The three strategies articulated earlier in this chapter all make sense at a theoretical level. Each provides a plausible set of directions by which the EPA can use the tools at its disposal to clean up inactive hazardous waste sites. Each addresses the contextual problem of limited administrative and financial resources. However, our aim in this research is to move beyond what makes sense in theory to an assessment of what works in practice. Effectiveness must be defined in terms of a set of goals against which the various approaches can be assessed. Unfortunately, the goals of the Superfund program are no more susceptible to easy articulation than are the potential implementation strategies to achieve those ends.

Superfund is not unusual in this regard: the goals of many public programs are inchoate. Statutory objectives typically are too broad and amorphous to be of much help in guiding day-to-day activity, or—when objectives are set down in more specific terms—in resolving conflicts among objectives.[33] The agency must establish its own priorities, its own definition of programmatic success. This enterprise is infrequently the subject of public debate; indeed, it may not even be discussed within the agency because of substantial internal disagreement over ends, means, and priorities. As a result, there is typically no clear articulation of the goals and trade-offs that underlie policy implementation. If articulated at all, goals are often fashioned and announced on an ad hoc, after-the-fact basis to provide justification for past agency performance.[34]

Superfund has a broad objective that raises little controversy—the cleanup of inactive hazardous waste sites. Under close analysis, this statutory mission reveals several potentially inconsistent underlying objectives. These cross-purposes underscore the need for more precise standards by which to assess the various approaches to implementation of the Superfund statute. Before we move to our discussion of specific cleanups illustrating different implementation strategies, we will briefly set out several measures of success in Superfund.[35]

Superfund is foremost a cleanup program aimed at eliminating immediate threats to the public health and the environment and—over the longer term—returning contaminated land to a usable condition. The first objective of the program is therefore *application of an appropriate remedy* to clean up a site. This seemingly transparent standard inevitably raises the thorny issue of what is meant by cleanup: how clean is clean?

Congress did not specify cleanliness levels for different types of dump sites and pollutants. Rather, it addressed this issue indirectly. The statutes specify that remedies selected by the EPA should be "protective of human health and the environment" and "cost effective," and that, to the maximum extent practicable, they should utilize "permanent" solutions.[36] Second, Congress required that the remedy be consistent with a National Contingency Plan to be drawn up by the EPA to provide guidance and uniformity in response actions. Finally, the 1986 amendments to the initial Comprehensive Environmental Response, Compensation, and Liability Act (CERCLA) statute required that remedies be in conformity with all "Applicable, or Relevant and Appropriate Requirements" (ARARs) established by federal, state, and local governments.

As a second program objective, the statutes envision that individual cleanups should be conducted (or at least paid for) primarily by the statutorily defined responsible parties, at the *minimum cost to taxpayers*.[37] The multibillion dollar Superfund embodied a substantial commitment of public funds to the national cleanup endeavor, but this was expected to be a revolving fund, used primarily in emergencies, or when no viable PRPs could be located. The expansive liability provisions of the statutes, and the coercive tools provided to the EPA to compel PRP compliance, are testimony to an unambiguous congressional intent concerning responsibility for most of the cost of site cleanups.

A third objective of the Superfund program arises from the agency's duty to minimize potential damage to the public health and the environment. This suggests not only that the final remedy be adequately protective, but also that the *cleanup should be accomplished expeditiously*. The emergency removal provisions of CERCLA are the clearest illustration of this statutory concern. The delays in cleanup characterizing the first years of Superfund lay behind many of SARA's subsequent provisions, particularly the statutory deadlines established for the EPA to achieve various program milestones and targets for aggregate remedial activities.[38]

A fourth evaluative standard is the *minimization of transaction costs*. We discuss transaction costs separately because they pose a different set of issues from the other costs involved in Superfund cleanups. As we define them here, transaction costs consist of those expenses that are directly related to determining issues of liability and in forcing the private parties to live up to their obligations under the statutes.[39]

These costs arise because Superfund is based on a liability scheme, ultimately enforced in court. They include the EPA's costs in conducting searches to identify responsible parties at a site, producing estimates of the volume and toxicity of the waste contributed by various parties, and negotiating remedial design and cleanup objectives with the PRPs. Legal costs, almost certainly the largest element in Superfund's transaction costs, are borne by all participants in the decision-making process.[40] One of the most persistent criticisms of the Superfund program has been the high cost of the legal and administrative processes that typically precede actual site cleanup.[41] One lawyer we interviewed suggested the intimate relationship of Superfund and legal fees in the pronouncement that "toxic waste has been good to me." EPA Administrator William Reilly has observed that there "have been far more people in three-piece suits than moonsuits on these projects."[42] Since many of SARA's provisions were designed to encourage negotiated, rather than litigated, resolution of disputes in Superfund cases, the reduction of transaction costs continues to be a criterion for evaluation of the program.[43]

In our reading on Superfund and in our discussions with practitioners and observers during our research, we encountered another potential objective of the program. Proponents of the prosecution approach in particular often suggest that Superfund provides a means to *punish* those individuals and businesses that have despoiled the environment, independent of a concern with site-specific cleanup activities.[44] This punishment-oriented approach to Superfund explains some of the actions and attitudes of EPA and Justice Department officials. It is a major component of the position toward Superfund advocated by vocal segments of the environmental community.[45]

We do not include punishment as an evaluative criterion for several reasons. First of all, it is not in the statutory framework, either explicitly or by implication. Congress has proven quite capable of using the punishment-oriented tools of the criminal law in the environmental area, as the criminal provisions of RCRA and the clean air and clean water statutes demonstrate. It chose not to do so in CERCLA. Superfund simply was not defined as a program about moral rights and wrongs.[46] Second, the retroactive nature of the liability provisions of CERCLA and SARA, particularly when combined with the judicial application of strict, joint and several liability, make punishment a particularly inappropriate objective in the Superfund scheme. Fundamental constitutional principles of due process and the prohibition of ex post facto legislation would all come into play in a statute explicitly aimed at punishing the parties who find themselves caught in Superfund's broadly cast net. It is unlikely that a criminal statute containing such extraordinarily broad liability provisions would survive constitutional scrutiny.[47] While some dump owners, operators, waste generators, and transporters knowingly engage in reprehensible activities, the range in blameworthiness within the class of Superfund PRPs, however this slippery concept is assessed, is substantial. The most culpable actions can be reached by the criminal provisions of other statutes. Responsible parties at the other end of the blameworthiness scale—such as unknowing purchasers of property later found to contain toxic substances, or generators who made good-faith efforts to dispose of their wastes responsibly—would seem to be peculiarly inappropriate subjects for punishment, especially in light of the harsh civil provisions of the statutes.

We thus have four evaluation criteria for assessing program success at a particular site. A successful cleanup should involve (1) an appropriate and cost-effective remedy, (2) at the least possible expense to the taxpayers, (3) obtained speedily, and (4) with minimal transaction costs. These criteria are obviously in tension. One would expect, for instance, that speed of cleanup and minimization of transaction costs must be traded off as a remedy increases in cost (and, presumably, protectiveness), and as the federal government's share of those costs decreases. Unsurprisingly, tensions in the objectives of the Superfund program have not been explicitly resolved by EPA; indeed, they are seldom publicly discussed.

Part 2 (chapters 3–5) provides case studies of the prosecution, accommodation, and public works approaches in action, at particular hazardous waste sites. The various evaluative standards set out above informs those discussions. A more systematic evaluation of the three implementation strategies in terms of these goals is provided in chapter 6.

Part Two
Case Studies

Three

Prosecution

THIS CHAPTER and the two that follow examine actual hazardous waste cleanups utilizing, respectively, the prosecution, accommodation, and public works strategies. Each chapter focuses on one EPA region—reflecting the distinctive regional orientations toward Superfund that track the rough typology of potential implementation strategies set out in chapter 2. We first discuss characteristics of the region and their relationship to the implementation strategy typically applied to Superfund cases. We then describe the events of two actual cleanups—one difficult and complex, the other comparatively simple. Finally, we use these cases as a means of examining the extent to which each strategy comports with its theoretical assumptions, and we assess in a preliminary fashion the degree to which the various goals of Superfund were satisfied. A more complete comparative evaluation of the strategies and the extent of their success at achieving programmatic goals is presented in chapter 6.

Regional Characteristics and Strategic Choices

The Environmental Protection Agency's Region V is headquartered in Chicago. It encompasses the states of the industrial Midwest: Minnesota, Wisconsin, Illinois, Indiana, Michigan, and Ohio. The region is seen in Washington and elsewhere as one of the two most consistent exemplars of a prosecution-oriented approach to Superfund implementation. (The other is Region II, based in New York City.) The image presented to the world by these offices is one of a tough, enforcement-oriented region that compels potentially responsible parties (PRPs) to clean up sites according to rigid schedules and coercive procedures. One former top EPA enforcement official said, "Some people think Region V is the toughest enforcement region: cut no deals, take no prisoners." Speaking from a different perspective, a former EPA deputy administrator told us that Region V "appears to be concerned about being seen as soft on industry."

The prosecution strategy used in Region V seems driven in part by the volume of business and the complexity of sites in the region, and in part by a set of more idiosyncratic administrative problems. The regional office is ultimately responsible for cleaning up more than 20 percent of the sites on the National Priorities List (NPL), a larger Superfund work load than that of any other EPA region. Many of these are typical "hard" sites: old industrial dumps involving multiple parties and poor records. Such sites inevitably require substantial administrative, legal, and scientific work before cleanup can commence. The sheer volume of effort entailed by the presence of dozens of such sites in the same region is magnified in Region V by complex and frequently strained relations between the EPA and state regulators, by chronic shortages of personnel, and by a high staff turnover, particularly at the critical remedial project manager (RPM) level.

The prosecution orientation provides a means of addressing personnel problems by transferring many of the tasks prerequisite to cleanup onto the PRPs. These tasks include determining the identities of all the PRPs involved with a site, determining how costs are to be allocated among those parties, conducting the requisite engineering studies, and designing and actually conducting the cleanup. If a small group of "deep-pocket" corporations or government entities can be forced to perform these activities, the responsibilities of the regional office correspondingly are reduced to issuing administrative orders, monitoring compliance, and carrying out enforcement actions against recalcitrants. In a region that is long on major Superfund sites and short on staff, particularly experienced site-level personnel, the appeal of this approach is obvious.

Other elements of the prosecution strategy are reflected in Region V: strict adherence to procedure, an emphasis on legal concerns, an aversion to forming informal relationships with PRPs and—especially—to being perceived as compromising the strong legal position of the government in Superfund cases. All of these aspects of Region V's prosecutorial strategy can be traced, at least in part, to the region's limited resources and heavy work load.

Regional leadership has created a structure of rigid procedures and schedules to guide and limit the discretion of the often inexperienced line staff. To conserve scarce supervisory resources and to enforce the image of implacability, Superfund supervisors in the region reportedly are reluctant to intervene in disputes between lower-level project staff and PRPs—thus allowing no avenue of appeal when PRPs feel they are being dealt with unreasonably by inexperienced or obdurate project managers.

Prosecutorial strategies show the stamp of a system dominated by lawyers and legal concerns. Again, regional characteristics and needs promote this orientation. Superfund leadership in Region V is exercised by a moderately stable and cohesive group of experienced managers, many of whom are lawyers. The prevalence of lawyers in regional management and at individual site negotiations derives from the underlying mandate of the prosecution strategy: if the government's strong legal position is to coerce PRPs into compliance, then most critical decisions will be legal in nature.[1]

The key legal element in the prosecution approach to Superfund is the doctrine of strict, joint and several liability. This notion of liability shapes thinking about the entire process of Superfund implementation. For example, regional officials often express antipathy toward mixed funding and other arrangements by which the government and PRPs share costs and risks at waste sites. An official in Region V's Office of Regional Counsel (ORC) told us, "We take the view that any one of the parties is liable for 100 percent." The regional position seems to be based on the query, Why should the government engage in mixed funding or risk-sharing when the potentially responsible parties are liable for everything? Indeed, one former official of the ORC told us, "A significant percentage of agency attorneys think that it is morally wrong to 'mixed-fund.'" A similar reluctance is exhibited toward de minimis buyouts, which provide full releases from liability to minimal contributors at a site in exchange for payment of a negotiated contribution to the cleanup. Such actions are seen as letting petty miscreants off the hook while requiring unnecessary administrative effort and multiple bureaucratic clearances.[2]

The continuing need for resources to meet a heavy work load has also affected the region's response to national work-load or productivity incentives in the Superfund program—the so-called bean count. According to one former EPA official with experience in both national and regional affairs, Region V is adept at the bean count game, performing well on the EPA's quarterly performance indicators and other benchmarks of achievement in order to reap additional staff resources.[3] The measurement scheme produces strong pressures to complete various administrative actions such as remedial investigations and feasibility studies (RI/FSs), records of decision (RODs), and remedial designs and remedial actions (RD/RAs) in time to meet the deadlines for quarterly reports to headquarters.[4]

Another countable bean is referral of a case to the Department of Justice for enforcement actions. The large number of Justice Department referrals from Region V is partially a result of the incentives created by this work-load measure, but referral to the Justice Department is also the centerpiece of a legalistic, prosecutorial approach to Superfund. The Justice Department referral process—with multiple stops within the region and at EPA national headquarters—also means that responses to PRP queries and requests are often slow in coming, and the multiple decisionmakers may not always produce consistent cues.[5]

The concern about meeting deadlines for reportable activities, in the context of a complex division of labor, a substantial work load, and a prosecutorial approach, also influences the degree to which Region V shares information and engages in consultative relationships with PRPs. Working under deadlines is hard on project managers, and even tougher on their supervisors, who frequently must make many decisions in the last few days of a reporting period. This situation can truncate the period for public comment on a proposed record of decision establishing the remedy to be selected at a site—often the only vehicle for PRP input into administrative decisions concerning site remediation in Region V.

Taken together, these practices are the source of Region V's reputation among PRPs as a large, impersonal bureaucracy. The RPMs who deal most directly with the PRPs are replaced frequently, and their behavior is constrained by established procedures. Informal input into decisions is discouraged by a secretive, bureaucratized system characterized by adversarial, distrustful relations between PRPs and the agency. In this context, it is easy to see how PRPs get the impression that they are dealt with "by the numbers," in a perfunctory and cavalier, if not hostile, manner.[6]

The two cases we examined in Region V illustrate many of its tendencies toward bureaucratic behavior. The first, Laskin Poplar, was a complex case involving hundreds of PRPs and cleanup costs that approached $25 million. In contrast, Cliffs Dow was a relatively simple case with a handful of PRPs and a comparatively low predicted cleanup cost. We have seen that the prosecutorial strategy in Region V is rooted not only in a commitment to the basic theory of that approach, but also in a set of organizational and practical needs operating at the regional level. The primary purpose of this chapter is to show how that strategy worked in practice.

Laskin Poplar

The Laskin Poplar site is on a nine-acre plot of land in Jefferson, Ohio. The site had been used since the turn of the century for greenhouse operations. In the 1950s, oil-fired boilers were installed to heat the greenhouses and Alvin Laskin, the owner and operator, bought waste oil to use as fuel. When the nursery business flagged, Laskin continued to purchase waste oil, selling it for a variety of purposes, including dust control.[7]

From at least the mid-1970s, Laskin's neighbors had complained periodically about odors emanating from his operations. Township officials finally took these complaints to the Ohio Environmental Protection Agency. The Ohio EPA, in its investigation during the late 1970s, confirmed the existence of odors and found polychlorinated biphenyls (PCBs) at the site. About the same time, the federal EPA undertook administrative actions against Laskin for discharging contaminants without a permit, a violation of the Clean Water Act.[8]

The environmental problems at the Laskin Poplar site resulted from the improper storage, handling, and combustion of waste oils. Contaminated oil had been mixed with other liquids and either stored in tanks or dumped in ponds. Improper burning had produced dioxin and other contaminants. Releases from the tanks and the liquid and sludge in the ponds contaminated the surrounding soil and posed a threat to the groundwater. The contaminants included PCBs, dioxin, and lead.

Summary of Key Events

The history of the cleanup at the Laskin Poplar site is shaped by the EPA's repeated efforts to use its liability doctrine to force a small group of PRPs into compliance. The agency relied primarily on court actions and took an uncompromising attitude in its negotiations. Despite the EPA's dogged efforts, the PRPs engaged in extensive and time-consuming legal maneuvers with the government, and initiated a flood of litigation among themselves and against a host of smaller PRPs. What the EPA had hoped would be the path to a quick, decisive settlement focused on a few parties expanded instead into a slow, expensive, and acrimonious process.

BEGINNINGS. In the late 1970s, both Ohio EPA and the U.S. EPA began cleanup efforts at the site by proceeding against Alvin Laskin, a man who by most accounts was both a difficult person and something of a country mechanical genius. The agencies wanted Laskin to cease operations, to remove the contaminated oil, and to begin cleaning up the pollution on his land. Laskin chased investigators off his property (reportedly with threats of physical violence), ignored court decisions, and generally resisted government efforts to compel a cleanup. Although negotiations with Laskin failed, the Ohio EPA and its federal counterpart were successful in separate court actions against him. In 1980, an Ohio court ordered Laskin to shut down operations; a receiver was appointed pursuant to this order. In 1981, the EPA succeeded in federal court in getting Laskin fined and extracted a promise from him to undertake a series of cleanup actions.

Laskin continued to operate when he could; he either did not or could not comply fully with any of the court orders against him. His comparatively meager resources made him effectively "judgment proof."[9] By 1981, community dissatisfaction had mounted and concern was being expressed about the slow pace of the cleanup. Senator Howard Metzenbaum of Ohio was particularly critical of the EPA and its administrator at that time, Anne Gorsuch (later Burford), for not using the statutory powers granted by Congress to expedite the cleanup. Using its statutory authority under the Comprehensive Environmental Response, Compensation, and Liability Act (CERCLA), the EPA began work during 1982 and 1983 to put the Laskin Poplar site on the National Priorities List. The EPA, with Ohio's concurrence, became the lead agency in these actions.

Physical conditions at the site were deteriorating throughout this period, and the federal government took a series of unilateral actions to deal with pollution on the site in the early 1980s. The EPA, using its own funds and its emergency response powers, undertook a 1981 cleanup of waste water in two surface ponds and worked to prevent further discharges into nearby Cemetery Creek. The following year the agency used its emergency removal powers again to remove 310,000 gallons of waste oil, and to treat contaminated water on the site, remove sludge, and cover some of Laskin's storage pits.

In 1983 and 1984, the EPA hired consultants and completed the first phase of the feasibility study that was a prerequisite to the more substantial work needed to clean up the site. The quality of the engineering work on this study—done by a subcontractor—was disputed by some PRPs and reportedly contributed little to the final RI/FS. The government also decided during this period to divide the Laskin Poplar site into two operable units—one for source removal (getting rid of the pollutants currently on the site) and the other for final remediation.

By this point, the EPA was out of pocket for both the emergency removal costs and for the engineering work of consultants on the RI/FS. At the same time, it had become clear that Laskin did not have the financial resources to cover these costs or to complete the cleanup himself. The EPA therefore needed to find other parties to reimburse it for past costs and to pay the substantial anticipated future costs of cleaning up the site. CERCLA's liability provisions gave the EPA the authority to proceed against the firms that had sold waste oil to Laskin over the years. Unfortunately, Laskin's record-keeping paralleled the idiosyncratic nature of his other activities: the primary records of waste oil purchases consisted of scribbled notations on canceled checks. Later, when PCBs were found at the site, Laskin contended he had no idea who had sold him "bad oil."

Region V settled on a strategy of proceeding against a group of fewer than a dozen PRPs on whom they had the best information. Laskin, his businesses, and a creditor (who ill-advisedly pressed a claim to Laskin's property) were the original defendants;[10] they were subsequently joined by seven large corporations called the "New Defendants."[11] This group would bear the brunt of the legal action that followed, despite the clear fact that a far greater number of PRPs were involved with the site. The New Defendants also formed the core of what became known as the Laskin Task Group.[12] While preliminary discussions and negotiations between the EPA and these PRPs had occurred sporadically during the early 1980s, the federal government apparently found the prospects for negotiated settlement dim. In 1984 it issued the first of four section 106 orders against the Laskin Task Group, this one ordering the corporations to remove a quarter-million gallons of waste oil and waste water from the site.

About the same time, the federal government filed suit in federal district court seeking a declaratory judgment under CERCLA to hold the defendants responsible for past costs at the site (about $2.3 million for emergency actions and for the completion of the RI/FS) and for unspecified future costs. In this court action, the government hoped to act quickly and to focus its limited legal and administrative resources on a small number of viable PRPs, purposely including a number of large corporations with the capacity to pay.

Once the federal government succeeded in getting those viable PRPs held liable for past costs and obtaining a declaratory judgment on future liability, this group could be made to perform the work, or at least to pay for it. It would then be up to the PRPs sued by the government to seek reimbursement from the remaining PRPs in subsequent contribution actions in federal court. By a swift move, then, the government hoped to recover its past costs and place responsibility for all subsequent work squarely on a limited and manageable group. A similar targeting strategy was involved in the subsequent issuance of the section 106 administrative orders.[13]

Throughout the case, Judge Alvin Krenzler, the presiding federal district judge, proved reluctant to make his court a battleground. Rather, he repeatedly encouraged, cajoled, and pressured the parties to negotiate. At one point, Judge Krenzler contrasted the federal government's desire for a declaratory judgment in the case with the PRPs' preference for a comprehensive settlement that included those whom the government had not yet proceeded against. He said the government wanted a "fast track on a simple issue of a complex case," while the defendants wanted a "comprehensive slow boat to China." His preference, he repeatedly made clear, was for something "in be-tween."[14]

During 1985, the parties talked and offers were made and rejected. The defendants offered to pay the federal government half of past costs in return for a release from all future costs; not surprisingly, the government found this, and related offers, unacceptable. The government pressed for a trial on its request for a declaratory judgment to settle the issue of liability. The judge and the defendants resisted.

The federal government also began negotiations at this time with a group of additional PRPs—numbering more than two hundred by 1985—that it had not sued. The size of this group increased gradually. Most of these PRPs initially were identified on the basis of Laskin's spotty records; more detailed information came from discovery actions of various parties in the suits and countersuits in Judge Krenzler's court and through government use of administrative orders to produce documents. These negotiations with unsued PRPs paralleled those between the EPA and the sued defendants: the PRPs made promises of money, in amounts considered inadequate by the government, if they could be released from future liability or if caps would be placed on future cleanup costs.

NEW TACTICS. In 1985 and 1986, as negotiations were proving unsuccessful, both the New Defendants and the federal government changed tactics. The New Defendants initiated third-party actions against 220 of the PRPs not sued by the government. Through subsequent waves of third-party suits, that number grew to more than six hundred. These newly sued PRPs became known as the Third-Party Defendants and were a varied lot that included a number of large corporations, as well as small- and medium-sized businesses. A subset of this group argued that they should be excluded from liability altogether because their contributions of waste fell under the Petroleum Exclusion Clause of CERCLA; they henceforth became known as the Petroleum Exclusion Group.[15]

The volume of legal activity increased dramatically as the new participants proceeded with pleadings and discovery motions. Complicating matters further, the New Defendants later sued one another (in 1987) after the federal government contended in court that private parties cannot use the doctrine of strict, joint and several liability in third-party actions. This process of suit and countersuit occupied the period from mid-1986 to early 1989. At one point the volume of Laskin Poplar-related litigation was so great that it required more lawyers than were available in Cleveland to handle the cases.[16]

While the New Defendants (the initial deep-pocket PRPs) were attempting to spread the liability through third-party contribution actions, the federal government modified its approach to the case. In 1986, the government abandoned its attempt to obtain future costs through court action and amended the complaint to include only the claim for past costs.[17] At the same time, the government developed a piecemeal approach of utilizing a series of administrative orders to get PRPs to assume responsibility for discrete cleanup tasks. The first administrative order, issued in 1984, dealt with the immediate threat posed by 250,000 gallons of waste water at the site. Subsequent orders, by contrast, addressed larger, longer-term problems. In 1986, a second order directed PRPs to develop work plans to clean up the storage pits, tanks and their contents, and soil surrounding Laskin's pits and tanks. In 1988, a third section 106 action ordered the PRPs to incinerate materials in the pits and tanks, as well as a portion of the most heavily contaminated soil. As indicated earlier, the EPA had already made the decision to segment work at the site into two operable units. Through these actions the EPA proceeded to subdivide the tasks still further.

While the federal government may have initially hoped for a comprehensive solution at the Laskin Poplar site, with the work done by a willing group of PRPs, the segmentation of tasks reflected the government's desire to move ahead in what was developing into a more fragmented environment. Laying the foundation for administrative orders, and for a legal defense of those orders should that prove necessary, involved much work, however. Involved were supervising consultants (who laid the technical foundations for the orders in conducting the work for the RI/FSs and other studies), providing and sometimes developing the documentation for the administrative record required for each order, and providing for and responding to public and PRP comments on the actions required.

The administrative record shows a mixture of adversarial and cooperative relations between the major PRPs and the EPA. There were numerous letters from PRP lawyers and reports of conversations in which PRPs complained about their lack of access to EPA documents, drafts of reports, and other information. In response, the agency repeatedly insisted that administrative decisions were its business, and that the PRPs would be given opportunities to comment at the time provided by the relevant statutes. The EPA's remedial project managers, however, reported a more cooperative relationship concerning engineering aspects of the site-cleanup activities being conducted by the Laskin Task Group's managers, consultants, and contractors.

FINAL NEGOTIATIONS. From 1985 through 1988, the EPA was involved simultaneously in several complicated negotiations and processes:

—A suit in federal court for reimbursement of past cleanup and response costs against the original and new defendants.

—A series of section 106 administrative orders directed at an expanding group of the larger PRPs. Each of these orders required the technical foundations to be laid by the work of consultants, and negotiations with state and federal authorities as well as the named PRPs. Once administrative orders were handed down, the EPA had to monitor compliance with those orders, a process which again involved it in complicated discussions with the same parties.

—Participation as an interested party in the third-party litigation being conducted by the PRPs. EPA and Justice Department lawyers were quite concerned about giving up legal ground. For example, they expressed disagreement with the New Defendants' contention that these defendants had the doctrine of strict, joint and several liability available to them in their actions against the third-party defendants; the government contended that the doctrine was available to it alone.

—Negotiating with PRPs over offers and counteroffers for settlement of past claims, future costs, or both. Included in this were separate negotiations with the New Defendants, the Petroleum Exclusion Group, and the ad hoc groups of settlers who periodically formed to extend their collective offers. Issues included whether or not the government would participate in mixed funding, the threshold levels and premiums for de minimis buyouts, the adequacy of dollar sums proposed, and promises to be made concerning future legal action.

Complicating the EPA's position further were its relations with the State of Ohio. Although Ohio EPA had rejected an initial suggestion that it take charge of the site, state environmental officials nonetheless had gotten involved. They had definite preferences about how

studies should be conducted, what should be required of PRP contractors, which state regulations should be applied at the site, and
how protective the selected remedial actions should be. Because
many of their preferences were channeled through Region V officials,
dealings with the state added to the EPA's burdens.

Although the arenas of negotiation were disparate, the EPA's position throughout remained constant. The agency repeatedly insisted
that any settlement cover the full amount of past costs, and that the
PRPs promise to pay all the (undetermined) future costs of the
cleanup. The desire for a comprehensive settlement also meant that
de minimis parties would only be able to buy out of the case as part of
such an all-inclusive settlement, a position also held by the major parties.

The initiative in organizing negotiations was borne by the big (or
what came to be called "*de maximus*") PRPs that formed the core of the
Laskin Task Group. One participant characterized the negotiations as
follows: "The PRPs fell out along a couple of dimensions . . . . [First],
the *de maximus* PRPs met a threshold contribution level (in gallons)
. . . and second, it was clear that these were mostly sophisticated
players with Superfund experience." This was, of course, what the
EPA had hoped would happen five years before.

In early 1989, the EPA and 146 PRPs (including the New Defendants) agreed to a consent decree covering approximately 80 percent
of past costs, leaving the government to proceed in court against the
nonsettlers for the remaining $2.0–$2.3 million. In late 1989, a second
consent decree was concluded to resolve the much larger issue of responsibility for the actual cleanup. In it the Laskin Site Group, the
twenty-seven companies that sent 7,000 gallons or more to the site
(this group was described in the consent decree as the *de maximus*
parties) agreed to do the cleanup, and 119 *de minimis* settlers agreed
to make fixed-dollar contributions to discharge their responsibilities.
The Petroleum Exclusion Group remained as nonsettlers.

Analysis

Region V's approach to the Laskin Poplar case provides a definitional example of the prosecution strategy. The region's aggressive
use of the coercive legal and administrative powers granted by CER-
CLA to obtain PRP compliance was evident throughout; conciliation

or accommodation with the PRPs does not appear to have been seriously discussed at any stage. Rather, the PRPs were consistently approached as "amoral calculators" in need of clear messages about the punishment that would follow a refusal to comply fully with the agency's demands. The central tenets of the prosecution strategy were consistently followed: quick and aggressive use of the agency's coercive legal and administrative tools, especially section 106 orders, minimal informal contact with PRPs, maintenance of a "hard-line" stance toward concessions.

The agency expected the prosecution strategy to transfer a substantial portion of the administrative and cleanup costs to a relatively small group of PRPs against whom the EPA had good evidence. These PRPs were then to take on the administrative and financial burdens of managing the cleanup according to standards laid down and enforced by the agency. Supplementing this approach was the use of administrative orders either to achieve goals that were not attainable in court or to put pressure on the PRPs for settlement.

The tendency throughout the case was to go by the book, both in following established routines and in guarding the government's legal and administrative prerogatives. The general complaints we heard at the beginning of our research concerning Region V's unwillingness to share information was echoed in complaints from attorneys in this case concerning their inability to get access to working drafts of studies until they were presented in the formal public comment period, after the basic engineering decisions had been made. The legal strategy followed in the case was also marked by an insistence on the government's legal prerogatives under joint and several liability. Even though the government ultimately settled for less than total past costs, its attorneys insisted that these were merely deferrals, rather than compromises of the principle of full reimbursement, since they would seek the remainder from nonsettlers.

In both the litigation and administrative-order approaches, Region V took the position that the burden of organizing a settlement agreement rested with the small set of targeted PRPs. Regional officials would pass judgment on settlement offers, but would not assume the role of initiator or mediator in the negotiations. The government chose instead to assemble the ingredients and shape the conditions under which a comprehensive settlement could occur: it put a limited number of PRPs on the spot, forced the potential *de minimis* parties to stay at the table by refusing to accept their individual buyouts offers, and made clear its requirements for a comprehensive settlement by repeated rejections of partial offers.[18] The government's role was therefore confined to that of a separate party that would only give final approval to acceptable proposals from the PRPs. A more active role would have required more ongoing attention and effort, requiring activity that Region V was ill-equipped to undertake.

Resolution of the case was complicated by the role played by the State of Ohio. Ohio, like many of the states in the region, has an active environmental agency: the Ohio Environmental Protection Agency. The Ohio EPA conducted many actions parallel to those of its federal counterpart: the state agency conducted initial studies at the site, litigated through its own attorney general's office, critiqued the work and work plans of the Laskin Task Group, determined which state regulations should be applied to work at the site, advanced its own preferred alternative for site remediation, and decided whether or not it would be a party to the resulting consent decrees.[19]

The parallel activities of the Ohio EPA created the potential for friction between the state and federal agencies, and between the Ohio EPA and the PRPs. Although the two environmental agencies performed similar functions for their respective governments, they worked from perspectives arising out of dissimilar political situations, responsibilities, and legal obligations. For these reasons, the Ohio EPA often took different positions from the EPA on the same questions: its studies produced different estimates of the extent of contamination at the site; it obtained an opinion from the state attorney general's office that differed with the federal government over the degree to which Laskin should be required to exhaust his assets before Ohio's contribution to emergency action would be considered; it pressed the EPA for a comprehensive, rather than piecemeal, approach to site remediation; and it expressed a preference for a more protective remedy than that tentatively favored by the EPA. Although the Ohio EPA did not always press these differences, substantial time and effort were required to resolve them. Ohio's demand that PRPs meet state cleanup standards that had not even been finalized was particularly offensive to the PRPs.[20]

The case of Laskin Poplar took more than a decade to resolve. Indeed, at this writing litigation is still pending against the nonsettlers and more lawsuits between responsible parties and their insurers are expected. The site conditions and general circumstances made it an archetypal hard case: a large site with a predictably expensive cleanup bill, an uncooperative operator, multiple PRPs, poor records, thorny legal issues, changing or developing state and federal requirements, and unsatisfactory initial work by consultants. Under these circumstances, speedy and inexpensive resolution of the issues was probably beyond the reach of any implementation strategy. There were, however, decisions under the control of the EPA and Department of Justice officials that might have been made differently.

Clearly, minimum levels of reliable information about individual contributions to the dump site must exist before meaningful settlement discussions can take place; some empirical information is essential for PRPs to determine both the existence of their liability and their proportionate share of the total cleanup bill. In the Laskin case, this information was generated during discovery procedures in the litigation and through government requests for information under CERCLA. A more direct and far less expensive method of building this information base would have been for the government to conduct a thorough PRP search before undertaking legal action.[21] But administrative gathering and processing of such information presupposes an organizational capacity that was probably not present in the Region V office. And such unilateral assumption of processing costs runs counter to the primary ethic of the prosecution approach: make the PRPs pay for the entire cost of the cleanup, including the costs of sorting out the identity and shares of the various waste contributors.

Next, all settlers wanted a sense that they were settling for relatively known or knowable costs. Unsurprisingly, the government prefers a broad performance standard rather than a fixed-price contribution from the PRPs because under the former the risks of design errors, faulty cost estimates, or other unforeseen circumstances are borne by the PRPs. The potential for virtually open-ended future costs, however, daunts corporate PRPs contemplating settlement.

In the Laskin Poplar case, the segmentation of work into operable units and the use of administrative orders probably provided some measure of predictability. The initial cleanup activities, mandated by the various administrative orders, simultaneously pared down uncertainty by reducing the number of remaining tasks and provided a better sense of what remained to be done. At the same time, the members of the Laskin Site Group were increasingly out of pocket themselves, which probably helped fuel their desire for a settlement under which they could spread their own burdens.

The unexplored alternative was for the government to have assumed some of the future risk by accepting fixed dollar sums from *de maximus* settlers for potential remediation costs at the site. Our interviews with PRP attorneys suggest that most PRPs are willing to pay a significant premium over what they might accept in a traditional open-ended consent decree if they can obtain certainty that this amount has bought them closure in the case, and freedom from the possibility of future costs arising from it. Assessment of such a premium in exchange for a reduction of future risks would have made settlement more attractive to the PRPs. Because such a decision could have been reached earlier in the process, there is reason to believe that it would have reduced the delays and high transaction costs that characterized the Laskin Poplar case.

Underlying the EPA's insistence on making the targeted PRPs bear all the costs and uncertainties of settlement was the notion that the government had an unassailable legal case. Governmental assumption of information-gathering and processing costs, and government assumption of any of the risks of unexpected events at the site, undoubtedly seemed poor uses of the program's limited administrative and financial capacities, and a squandering of the strength of its legal position. But the calculations might have come out differently had the government been able to anticipate the reticence of the federal judge to bring the issue to trial on the issues the government presented, and the difficulties inherent in disposing of what government attorneys considered to be the peripheral claims and counterclaims in the case. Liability doctrines may be entirely on the government's side, but few judges welcome the potentially lengthy (and, from their point of view, boring and unproductive) trials involved in proving them. Furthermore, in jury trials there is always a certain amount of danger to the government in even the appearance of unfairness or unreasonableness in its dealings with private parties, regardless of the strength of its legal position.

Cliffs Dow

The Cliffs Dow case differs from Laskin Poplar on a number of dimensions: expected remediation costs, number and cooperativeness of PRPs, and complexity of the scientific and technical issues involved in the cleanup. It provides an opportunity to examine how the prosecution strategy functions at a site seemingly more susceptible to a speedy, less contentious process of decisionmaking.

The Cliffs Dow site occupies about two acres of land covered by woods and scrub on the outskirts of Marquette, Michigan. Although the site is on low ground, it is close enough to the Dead River and to Lake Superior to raise concern that these may be affected by the site.

The City of Marquette owned the land, leasing it from 1954 through the 1960s to the Cliffs Dow Company. Cliffs Dow was initially owned jointly by Dow Chemical and the Cleveland Cliffs Iron Company. It was sold in 1968 to the E.L. Bruce Company, which did business under the name of Royal Oak Charcoal. The site had been used to dispose of wood tars from the manufacturing of charcoal over the period of the lease arrangement. Together, the companies disposed of 8,000–20,000 cubic yards of wastes at the site. The EPA listed waste types at the site as solvents, tars, and resins from charcoal processing.[22] Soil contamination was found to a depth of ten to seventeen feet.

Summary of Key Events

Region V took a prosecution-oriented approach to the parties at Cliffs Dow: issuing the requisite notices, providing the formal periods for response, and otherwise indicating that the full apparatus of the Superfund liability doctrine would be brought to bear should the parties not cooperate. In contrast to the Laskin Poplar PRPs, however, most of the Cliffs Dow parties chose to cooperate. The EPA subsequently withdrew to a role of monitor to insure that PRP-conducted activities met agency requirements.

The case began in the spring of 1981, when two hikers complained to the City of Marquette that their clothing had been soiled by tar residue while they had walked on the site. The city initiated an investigation and later referred the issue to the EPA,[23] which ultimately placed the site on the National Priorities List.

There was little question concerning the identity of the responsible parties, since the site was owned by the City of Marquette and had been used as a dump by a specific set of companies operating the nearby charcoal manufacturing plant. The contaminants were the unmistakable residue of those activities. While contaminants had soaked into the ground, the terrain prevented it from migrating off site. Thus, no emergency removal was required.

The EPA issued a special notice letter to the only PRPs identified at the site: the City of Marquette, Cliffs Dow Chemical (and its parent companies), and the E.L. Bruce Company. The letter offered the PRPs an opportunity to conduct the remedial investigation/feasibility study of the site. Once notified by the EPA, the PRPs held discussions among themselves. All except the E.L. Bruce Company decided to conduct the RI/FS study required by the EPA, and their agreement was formalized in an administrative order by consent. Ultimately, the companies spent $1 million on this effort, hiring engineering consultants and forming a task force to coordinate the study.[24]

From 1984 until early 1988 work progressed at the site and on the RI/FS. There were several disputes over the RI/FS, however: the EPA faulted the PRPs for the slow pace of their work, for what the agency saw as technical deficiencies in the scientific studies, and for alleged failure to deal properly with applicable or appropriate relevant standards (ARARs)—the pollution standards of other government units that the cleanup ultimately must satisfy. The PRPs contended that the EPA was asking them—through the ARARs and other means—to address problems not anticipated in the original administrative order.

The disputes over the RI/FS continued throughout 1988. At one point the EPA threatened to take over the study itself, but the agency backed down and ultimately accepted the disputed portion of the feasibility study. In making their final determination as to remediation, however, EPA officials announced that they would not limit themselves to the remedies described in the feasibility study. Instead, they created an expanded list of remedies dealing with problems they thought ought to be addressed at the site.

As the EPA was preparing to issue a record of decision based on its expanded list of alternatives, the PRPs filed a motion in federal court for a temporary restraining order to prevent the EPA from considering alternatives not studied in the RI/FS. The court denied the motion, leaving the agency free to issue a record of decision containing a remedy not assessed in the feasibility study.

The dispute between the EPA and the PRPs centered on the agency's rejection of the alternative favored by the PRPs, and the agency's initial preference for a more expensive and more extensive remedy than was contemplated in the RI/FS. After a series of discussions and a period for public comment, the regional administrator selected a remedy that split the difference in costs between the PRPs' favored alternative and the remedy initially favored by the EPA. The record of decision was issued in September 1989. It required incineration of the most badly contaminated soil, bioremediation for the remainder, and a monitoring scheme to ensure that the contaminants were fully removed.

Faced with a completed record of decision and the prospect of an administrative order to do the work or face treble damages, the PRPs chose to design and conduct the remedial work on the site. Again, all PRPs except the E.L. Bruce Company participated. The work, according to one interviewee, cost the PRPs much more than they had anticipated. By the end of 1991, the soil incineration portion of the work had been completed. The bioremediation solution was planned for some of the remaining contamination. The case was in its final phases as research for this book was being completed. The remaining bones of contention concerned allocation of monitoring costs and quality standards for groundwater at the site. The latter issue was particularly irritating to the PRPs. As in Laskin Poplar, state officials had demanded that the remedy be consistent with water standards that had not yet been announced, or even formulated.[25]

Analysis

Despite differences in size and complexity, both Cliffs Dow and Laskin Poplar illustrate most aspects of the prosecution strategy: the willingness of regional officials to make aggressive use of enforcement powers, mutual distrust exhibited by the agency and the PRPs, and the tendency of government officials to operate according to established routines.[26] These characteristics are manifested in the sequence of events and in the administrative structures and procedures within which the case was handled.

The active and cooperative participation of PRPs from the outset distinguish Cliffs Dow from Laskin Poplar. Our informants in Region V suggested that the PRP-conducted RI/FS at Cliffs Dow was something of an anomaly in Region V, an assertion that is supported by aggregate data.[27] Regional officials told us that they rarely encourage PRPs to do this work because of a preference to control the process themselves. Indeed, in Laskin Poplar and other cases in which the region conducted the RI/FS, the EPA jealously guarded the information produced until the officially designated comment periods. In this case, a PRP-conducted RI/FS constrained Region V's preferred operating style by introducing another participant into the process at an earlier stage. The result was controversy over both the scope of problems and the range of remedies. Characteristically, however, the region threatened to take over the study and later used an administrative procedure to obtain what it could not get through oversight of the PRP-conducted study.

Michigan's actions in the Cliffs Dow case were roughly analogous to the role played by Ohio in Laskin Poplar. Michigan has an active and aggressive environmental regulatory structure. As in Laskin Poplar, the state allowed the EPA to take the lead at the site. State officials could therefore observe and critique activities of the EPA and the PRPs without assuming any responsibility for their success. Because state officials had no financial stake in the final cost of the remedy at Cliffs Dow, they could demand high levels of cleanliness. As a further complication arising from the PRP conduct of the cleanup, negotiations concerning cleanup levels were frequently conducted directly between the PRPs and the state, with no federal intervention. In such circumstances, the PRPs were placed at a disadvantage, since state officials made a changing set of unilateral demands concerning AR-ARs and final cleanliness levels—several of which (as in Laskin Poplar) had not even been formulated when the remedy was being designed and implemented.

The context was extremely favorable for settlement in Cliffs Dow: clear liability, solvent PRPs, an isolated site posing few dangers to neighbors, and comparatively modest projected cleanup costs. Given these ingredients, it is not surprising that PRPs chose to take a cooperative stance. In many ways Cliffs Dow was a perfect site for a cooperative strategy based on mutual adaptation. The EPA wanted the work done and the PRPs had a clear stake in being cooperative. While the interests of participants were not identical, their reasoned preferences were not far apart. Under these circumstances, the level of acrimony that ultimately developed in the case is puzzling.

The complexity and ambiguity of Superfund's cleanup requirements, and the attendant uncertainty about what had to be done, helped introduce conflict into what might have been a relatively smooth process of mutual accommodation. This tendency was heightened substantially by an implementation strategy which holds that all PRPs are "bad guys" who must be coerced into doing the right thing—despite their protestations and actions to the contrary—and which features a style of dealing with private parties that the cooperating PRPs must have regarded as more appropriate to a district attorney's office. While Region V's behavior may be understandable in light of its administrative needs, resource constraints, and pressures from the state, the piecemeal revelation of new requirements over time exacerbated tensions. Also, the aggressiveness that sometimes characterized agency actions toward the PRPs seemed gratuitous in light of their consistent efforts to be cooperative. The agency's last-moment flexibility in remedy selection did help bring about settlement, but it is unclear why this flexibility could not have been manifested earlier in the process.

Additional delays and costs were incurred by the EPA's use of the oversight process. Had the scope of activities, and some relatively comprehensive accounting of relevant cleanup criteria, been laid out at the outset, the PRPs would have had a clearer idea of what they were getting into. The appearance of arbitrary and rigid behavior, and the year of dispute over the feasibility study, might have been avoided. This would have required, however, a considerably greater initial investment by the EPA in specifying the requirements in the administrative order by consent.

Government strategy in most contexts is aided by a citizenry that takes its legal obligations seriously. Yet, in the context of this case, the PRPs' attempt to establish their good citizenship was an exercise in near-futility. As a practical matter, the routinized approach taken by Region V does not leave much room for recognition of past compliance. Given the turnover of remedial project managers and the accompanying absence of institutional memory, it is not surprising that personal relationships of trust did not evolve. While the cast of government actors was constantly changing, the PRP task force remained remarkably stable. This gave the PRP side an impression that it was dealing with a revolving-door bureaucracy that was constantly creating new requirements. This impression may have been derived less from changing requirements than from the regional practice (if not necessity) of dealing with problems only as they occurred. But the resulting image was the same: governmental arbitrariness, if not irrationality.

The EPA's strategy resulted in both the PRPs' agreement to conduct the RI/FS, and ultimately their willingness to undertake and pay for the cleanup. It was not a strategy driven primarily by cooperation, however. Rather, the agency derived maximum benefit from a situation in which the PRPs were clearly liable, in which they had a stake in being "good citizens," and in which they knew that, ultimately, they would have to pay for the work.

Assessing the Prosecution Strategy

The prosecution strategy is based on the assumption that targeted PRPs will take full responsibility for site cleanup if they are faced with a superior legal position, nonnegotiable demands, and the threat of sufficiently disagreeable consequences following noncompliance. Burdens on the government are substantially reduced by requiring targeted PRPs to finance, if not conduct, much of the work involved in site cleanup. The EPA's role in this scenario is limited to threatening sanctions for nonperformance, creating and enforcing schedules for completing tasks, and monitoring PRP activities at each step so that the requirements of the law are fulfilled. How were the EPA's expectations met in practice, and how did the results of these cases measure up against the various goals of the Superfund program?

There is little question that Region V personnel behaved in accordance with the dictates of the prosecution model in Laskin Poplar and Cliffs Dow. They relied heavily on strict, joint and several liability; they presented an image of implacability, as well as procedural and substantive rigidity, to the PRPs; they resisted negotiation—particularly in the early and middle stages of remedial decisionmaking. Even after concessions were made to obtain agreements, EPA officials insisted on calling them mere postponements of the complete and total victory which they considered the government's rightful reward in Superfund cases.

It was in the actions of PRPs that the prosecution strategy in practice deviated from its theoretical assumptions. While government officials behaved in accordance with the prescribed rules of prosecutorial behavior, PRPs did not become the submissive, compliant parties that a pure rational-actor model might predict. In the case of Laskin Poplar, the PRPs used delaying tactics, initiated scores of third-party suits, and banded together in opposition to what was perceived as government unreasonableness. Ultimately the PRPs succumbed to the superior force of the government's legal position, but not without some strategic successes along the way. In the case of Cliffs Dow, the PRPs were more cooperative. But in both cases the government's victory was not the complete triumph over the opposition that partisans of the prosecution strategy would suggest. And the partial victory in Laskin Poplar came at substantial cost—in terms of delay, inflated transaction costs, and the generation of ill will on the part of most participants in the process.

To what extent did the prosecution strategy, as exemplified in these cases, achieve the goals of Superfund set out in the previous chapter? It is difficult to answer this question outside of a comparative framework; when assessing the success of public programs, one should always ask, Compared to what? However, we can observe some outcomes against the expectations posited for the strategy at the outset.

Was the remedy appropriate? The EPA insisted on a host of stringent cleanup standards, but in the end compromised with PRPs. Alternatively, the PRPs initially favored less stringent and costly approaches than what they finally agreed to pay for. We cannot determine whether the final results were adequate or inadequate in an objective sense, although we saw no evidence of insufficiency on this score, but can say that the final decisions as to remedy were somewhere between the wishes of the PRPs and those of the government.

Did the strategy minimize taxpayers' costs? The settling PRPs did agree to pay for most of the future cleanup costs and a substantial share of the past costs in both cases. But the EPA also paid for some of the costs, particularly in the case of Laskin Poplar. The agency did so, however, with the expectation—or at least the hope—that it could recover some of those costs later from nonsettling parties.

Did cleanup proceed in a speedy fashion? The prosecution strategy almost certainly added delays to what was already an inherently difficult and lengthy process, especially in the case of Laskin Poplar. All parties would have benefited from speedier resolution of the issues since final PRP costs would have been reduced and the EPA would have saved time and organizational effort. But no single party was willing to sacrifice individual interests to obtain a speedy settlement.

Finally, there is the issue of transaction costs. In the case of Laskin Poplar this constituted a runaway problem, although it is reasonably clear that the contentiousness of the dealings in the case of Cliffs Dow caused substantial additional expense as well.

The foregoing suggests that the prosecution strategy achieved results in terms of only some of the goals of the Superfund program. This is hardly surprising in light of the inherent tension in the goals discussed in chapter 2. What the EPA tried to do, and what it achieved, was in fact inextricably tied to the strategy it applied. Our reading of the cases, however, indicates that there were some underused tools in the Superfund kit that might have contributed to a better, or at least a different, mix of results. Since the government compromised at the end, we can make conjectures as to what the effects of an earlier compromise might have been. There were undoubtedly important mutual interests—the benefits of speed and the minimization of transaction costs—that were submerged by the adversarial nature of the prosecution strategy. These too, under another approach, may have also contributed to different outcomes. In the next chapter, we explore the degree to which the accommodation strategy made somewhat different use of the mix of available policy implements.

Four

Accommodation

IN THE preliminary interviews we conducted in the Superfund community for this book, the impression we gained of
Region III was one of managerial competence applied to a strategy for
implementing Superfund that is especially successful in obtaining
settlements from potentially responsible parties. It is generally believed that these settlements are reached through an approach that is
tough on recalcitrant parties, but is frequently flexible and innovative
in dealings with cooperative PRPs. This approach seemed to be consistent with the accommodation strategy and we went to Region III to
find out how it worked.

Regional Characteristics and Strategic Choices

The Environmental Protection Agency's Region III, headquartered
in Philadelphia, covers the Middle Atlantic states of Delaware, Maryland, Pennsylvania, Virginia, and West Virginia. In comparison to the
programs of other EPA regions, the Superfund program in Region III
is large. With jurisdiction over 152 National Priorities List sites, the
office is responsible for more cleanup activity than all but three of the
EPA's ten regions.

Initial interviews in the Superfund community led us to characterize Region III as oriented toward negotiation and settlement. While
the management in all EPA regions is interested in settling rather than
litigating Superfund cases, it appears that Region III's approach to
PRPs is less confrontational than that of other regions, such as II and
V. We also heard that remedial project managers (RPMs) in Region III

are more likely to share technical information with PRPs, and to take their input seriously in decisions embodied in the remedial investigation and feasibility study (RI/FS), or in the choice of a remedy finalized in the record of decision (ROD). Further, we expected to find more use of the settlement devices embodied in the authorizing legislation: nonbinding allocations of responsibility, mixed funding, *de minimis* buyouts, and the like.

Generally, there seems to be an appreciation at the project-manager and regional-counsel level in Region III that many PRPs are interested in cooperating with the agency, and share the goal of getting sites cleaned up expeditiously. While these impressions are tempered by a realization that corporations are probably most concerned about minimizing their costs in a cleanup, and that some PRPs simply cannot be trusted, we sensed that regional EPA staff try to cooperate and to accommodate reasonable requests by PRPs, and that they appreciate some of the difficulties and constraints under which PRPs operate.

Region's III's adoption of the accommodation approach to Superfund implementation appears to be related to several factors: the work load and management style of the regional office, an aversion to litigation, the professional orientation of leadership in the office, and the generally cordial relations between the region and both EPA headquarters and state regulatory authorities.

When we conducted our interviews in Philadelphia, the regional office seemed to be relatively well-staffed and competently managed. We do not have data on regional staffing levels for the Superfund program, but regardless of the objective reality of regional work load, we did not sense in Region III the overpowering pressure of too much work and too few personnel that seemed to burden the Superfund program in Region V. Nor did we hear complaints about rapid staff turnover at the RPM or regional counsel level. Several interviewees observed that the management structure in the office was also quite stable. Frequently we found that middle- and upper-level managers in both the Waste Division and the Office of Regional Counsel (ORC) were "old-timers."

A stable, experienced group of project managers that is not overburdened with work can afford to spend more time in discussions and negotiations with PRPs and their engineering consultants, and need not be held in check by the strict adherence to procedure that may be obligatory with an overworked and less experienced staff. We saw no aversion to mixed funding, *de minimis* buyouts, or assistance with share allocation in Region III. Rather than the view that mixed funding was somehow immoral—a view expressed in Region V—several officials expressed pride in Region III's groundbreaking mixed-funding agreements.

The stability and size of staff, together with the wider discretion granted to program personnel, contributes to the generally high morale we observed (and others reported) in the regional office. We also noted in interviews within the region that its managers focus on organizational issues within the office: ways to retain good RPMs and other technical personnel, staff training, creating a "team" atmosphere.

Related aspects of the organization of the regional office also influence its use of the accommodation strategy in Superfund cases: the professional orientation of regional managers, and the relationship between the program and legal staffs. Unlike Region V, where lawyers often overpowered the program people, in Region III the program staff seem in control. The relatively low turnover may strengthen this tendency: as RPMs—often recent engineering and science graduates—become more experienced in managing Superfund cases, they undoubtedly have greater confidence in their judgment when challenged by lawyers.

However, a more conscious policy may be at work in Region III. The relationship of the ORC to the program staff in the Waste Division is particularly important in this regard. We heard from lawyers in the ORC, as well as program staff, that in Region III, the relationship between the ORC and the program divisions is one of "lawyer and client." One RPM said, "They [lawyers in the ORC] have a lot of influence over what happens in enforcement cases, but they ultimately have to defer to us." A lawyer in the ORC suggested how unusual he considered this relationship: "When the guys from DOJ [the Department of Justice] come up here, they go crazy; they can't believe we consider ourselves lawyers for 'clients' in the Superfund program." We believe the absence of program domination by lawyers in the regional counsel's office is a key factor in the region's use of an accommodation approach.

We also discovered in Region III a strong aversion to litigation, even within the ORC. It is unclear whether this bias is a cause or an effect of the region's orientation toward an accommodation strategy (we suspect it is both), but we frequently heard in interviews that the goal in Superfund cases should be to reach an amicable settlement rather than get involved with legal proceedings. One of our Region III case studies, Tybouts Corner Landfill, was a complex and contentious donnybrook that lasted more than a decade in federal court. It was something of a watershed experience for the region in this regard. One manager said, "Anytime somebody gets frustrated with the PRPs and suggests we go to court, all we have to say is, 'Remember Tybouts,' and he usually shuts up."

Finally, we suspect that an accommodation strategy is most plausible when the external political and administrative environment is supportive, or at least not hostile. Essential to the accommodation strategy are close, informal relations with PRPs and a willingness to negotiate and compromise. These activities involve risk, particularly in an environment in which such activities can be perceived as "making deals with polluters." The managerial successes of Region III have led to supportive relations with EPA headquarters in Washington. Like Region V, Region III is reportedly to be good at the bean count game, although we heard fewer comments about such activities and their impact on staff behavior in Philadelphia than we did in Chicago. The generally high regard in which the region is held nationally is indicated by the fact that Region III staff frequently lead various national programs and committees. The regional office also has served as a source of recruits for leadership positions at EPA headquarters. Such conditions tend to give regional staff the confidence to take the political and administrative risks inherent in a process of accommodation and compromise with PRPs.

Region III also appears to enjoy relatively cordial relations with the state governments in its region, an additional factor contributing to its flexibility in dealing with cleanups. Most of the region's Superfund activity is concentrated in Delaware and Pennsylvania. The relative lack of Superfund activity in Region III's other three states is reportedly a function of a limited number of sites in Maryland and Virginia, and a reluctance on the part of state officials to have federal environmental officials involved in site cleanups in Virginia and West Virginia. Of the two states in Region III with the most Superfund activity, Delaware has tended to keep a low profile in hazardous waste matters and generally relies on EPA efforts. This hands-off attitude is illustrated in both case studies, in which—unlike in Region V—the state was not a major player. Pennsylvania, by contrast, has a more aggressive regulatory bureaucracy and has adopted an adversarial stance toward regional EPA officials at some of the state's major Superfund sites. While none of our case studies involved Pennsylvania sites, we understand that the region's ability to exercise flexibility in negotiating with PRPs is more constrained there than elsewhere in Region III.

The two cases we examined in Region III were Tybouts Corner Landfill and Harvey and Knott Drum. Analysis of the Tybouts case as an instance of accommodation is complicated by the fact that attorneys from the Justice Department's Lands Division handled the case from Washington during the critical years leading up to its final settlement. That stage of the case is better characterized as prosecution, rather than accommodation.[1] On the engineering side, however, the region's relatively open RI/FS processes were demonstrated in Tybouts, as was the region's receptivity to innovative technology. The final consent decree in the case, despite the fact that it followed years of bickering among PRPs and between PRPs and the government, did embody one of the first mixed-funding settlements in the United States. Harvey and Knott Drum, a relatively simple site that was handled primarily within the region, is similarly illustrative of an approach to site cleanup in which the PRPs are treated more as negotiating partners than as miscreants who cannot be trusted and who are in need of punishment.

Tybouts Corner Landfill

The case of Tybouts Corner Landfill was marked by political, scientific, and legal controversy. Befitting a site that ultimately was ranked second among the twelve hundred sites on the NPL, its progress (and frequent lack thereof) was closely monitored by congressional committees, environmental activists, the national media, and local newspapers and citizens' groups. Tybouts Corner Landfill was a case in which substantial new ground was broken in the Superfund program. Its resolution included one of the first mixed-funding agreements negotiated by the EPA; indeed, Tybouts was one of the first cases in which PRPs agreed in a consent decree to clean up an NPL site. Given these factors, it is not surprising that resolution of the Tybouts case took so much time. The issues involved at the site were daunting:

—Allocation of responsibility for a multimillion-dollar cleanup among more than two dozen PRPs—some sued by the government, some brought in as third-party defendants by the original defendants, but all with differing degrees of involvement with the landfill;

—Determination of the degree of government participation in the cleanup through a mixed-funding agreement involving Superfund monies;

—Specification of the remedy for a landfill of nearly fifty acres, with a depth of five to forty feet, leaching several hazardous substances into surrounding groundwater and aquifers;

—Provision for the scientific and engineering uncertainties that inevitably accompany a remedy that will be operational into the twenty-first century.

Even the most pessimistic estimates at the commencement of the case probably would not have forecast the time and resources consumed in the resolution of these issues. From the time contamination at the site was first documented by the State of Delaware until the consent decrees involving the major PRPs were lodged in federal district court, the process consumed more than thirteen years.

The Tybouts Corner Landfill is located in New Castle County, Delaware, about ten miles south of Wilmington. The most populous county in Delaware, New Castle is highly industrialized, with major chemical, automobile, and other manufacturing plants located in or nearby the county. New Castle County also has more Superfund sites within its borders (eight) than many *states*. Placed on the National Priorities List in 1983, Tybouts Corner is a few miles from three other NPL sites: Army Creek, Delaware Sand and Gravel, and Harvey and Knott Drum.

The landfill was owned by William Ward, operator of a local gravel business. Ward bought the property in the mid-1960s as a source of gravel, and for its long-term investment potential. In late 1968 he was advised that New Castle County wished to use his property for a sanitary landfill, and that if he did not agree to sell the land to the county, it would be condemned. Because Ward wished to continue removing gravel and to retain ownership of the land for investment purposes, he negotiated an agreement in which he leased his property to New

Castle County. Ward agreed to operate the landfill with his employees and equipment. According to Ward, the county controlled the landfill gates and had sole authority over what was deposited in it.

The landfill operated from January 1969 until April 1971, at which time it reached capacity and was closed. The problems of pollution at the site were not revealed until May 1976, when the Delaware Department of Natural Resources and Environmental Conservation found organic compounds contaminating a private well 400 feet east of the landfill. About the same time, water in nearby Pigeon Run Creek and Red Lion Creek were also found to contain organic contaminants.

Summary of Key Events

The EPA filed suit against New Castle County, Ward, and the Stauffer Chemical Company in federal district court in Wilmington on October 8, 1980. The original summons and complaint were the first entry in a court docket that would run to more than two thousand entries before the end of 1989. Because the suit predated passage of the Comprehensive Environmental Response, Compensation, and Liability Act of 1980 (CERCLA), the suit was filed under the Resource Conservation and Recovery Act (RCRA) for the purpose of obtaining injunctive relief to abate endangerment to human life and the environment.[2]

The EPA's decision to proceed against only three defendants, at a site where a multiplicity of waste generators was evident almost from the beginning, was made in part because of a limited PRP search and in part because of a then-prevalent strategy of conserving agency resources by going after a few "slam-dunk" PRPs.[3] Ward and New Castle County were obvious candidates because of their respective ownership and operation of the dump. Stauffer Chemical, by its own admission in a previous congressional survey of hazardous waste disposal practices,[4] was unambiguously implicated as a waste-generator at the site.

While the lawsuit was getting underway, EPA technical staff began to address the concerns of the local residents, assess the degree of environmental damage, and examine possible remedies. After six months of negotiations with the PRPs over funding of the RI/FS, the EPA and the State of Delaware terminated discussions and agreed to do the study themselves.[5] The project was contracted out to an environmental consulting firm. Concurrent with commencement of the

RI/FS was a substantial increase in local political concern over the Tybouts dump. Front-page articles in Delaware papers appeared and acrimonious public meetings between the EPA and irate neighbors of the landfill became media events. The EPA's ranking of Tybouts as number two on the NPL on September 1, 1983, raised the political pressure another notch.[6]

The major issue motivating the local community during this period was the risk to residents whose wells were in danger of contamination from the plume of contaminants moving out from the dump.[7] EPA cost estimates for various solutions to the water-supply problem were not completed until almost a year later, in July 1984. The EPA ultimately decided to extend the county's municipal water supply to the forty-six residents whose wells were potentially at risk. The new water lines were completed in March 1985, through a cooperative arrangement in which Ward provided labor and materials at cost, and New Castle County and Stauffer Chemical compensated Ward for his out-of-pocket costs.

EARLY PROGRESS OF THE LAWSUIT. While the EPA was attempting to mollify residents and supervise conduct of the RI/FS, the lawsuit was making slow, although not necessarily steady, progress. The original three defendants were unhappy from the outset at being singled out by the EPA to bear full responsibility for a problem they believed had been caused by a far greater number of parties. Much of the early legal work on behalf of the defendants was directed toward documenting the complicity of other waste generators and trying to convince the EPA and the Department of Justice to add those parties as defendants in the action. These efforts bore only limited fruit: the government agreed to add ICI Americas, an international chemical company, to the suit in April 1984.

The addition of ICI Americas and its attorney to the suit, together with the August 1984 appointment of a new attorney for New Castle County, invigorated a lawsuit that had seen only halting progress during its four-year pendency. ICI, Stauffer, and New Castle County, through their respective attorneys, forged a close alliance with a common strategy toward most elements of the case. The strategy had two objectives: first, to involve as many additional deep-pocket waste generators as possible in the case, either as named defendants (the preferred option) or as third-party defendants in concurrent contribution actions; second, to influence the technical decisions being made by the EPA so as to minimize the total cost of the remedy.

Within two months of the entry of ICI Americas as a named defendant in the case, that company began a series of depositions of corporations that might be linked to the Tybouts site. In late April 1985, ICI (and, subsequently, Stauffer and New Castle County) acted on the information gained in this barrage of depositions and interrogatories by filing third-party contribution suits against more than twenty corporations. These actions provoked a storm of legal activity. Many of the third-party defendants filed counterclaims against the original defendants, and cross claims against other third-party defendants. (There was even an attempt by one of the third-party defendants to initiate a fourth-party suit.) These new parties to the litigation also began engaging in their own discovery in the case, attempting to play catch-up in a legal action that had been underway for five years.

The results of the legal maneuvering were as predictable as they were extraordinarily expensive for all involved. In the six months following the filing of ICI's third-party complaints, the court's docket records more than a hundred separate pleadings,[8] and more than 160 substantive discovery events.[9] The total number of docket entries in the case averaged out to more than three filings for each day the court was open during the period.

Evidentiary issues in the Tybouts case were complex because neither Ward nor New Castle County kept even rudimentary records from which a waste-in list could be derived. The nature, quantity, and generator of the material deposited at the landfill had to be determined inferentially, through painstaking depositions of trash haulers and employees at the dump. This evidence was then linked to parallel investigations of the manufacturing processes and waste streams for each product produced by the many industrial concerns that were potential users of the site. In an interview, a lawyer for one of the third-party defendants termed this expensive and intrusive process "scorched-earth" discovery.[10] Its scale and intensity generated substantial friction between the original defendants and those subsequently sued in third-party actions.

REACHING AGREEMENT ON THE REMEDY. Technical work continued concurrently with the litigation. The RI/FS was begun in

January 1983. While the initial stages of the remedial investigation were conducted without substantial input from the PRPs, the primary defendants (ICI, Stauffer, and New Castle County) became much more involved when specific alternatives began to be evaluated. By all accounts, the RI/FS process in Tybouts was open to the PRPs, both in terms of sharing the information being developed by the EPA's contractor as the soil and water samples were analyzed, and in terms of PRP input into the remedy being designed for the site.

The broad outlines of the Tybouts remedy became clear relatively early in the technical evaluation. The size, depth, and geological properties of the site suggested a remedy that would involve consolidating and then isolating the contaminants in the landfill. The PRPs proposed a remedy that differed from the traditional approach in two ways. First, they proposed that a network of subsurface drains be built around the fill, rather than the more usual slurry wall, to capture hazardous materials migrating from the site. Second, they proposed a modification in the design of the impermeable cap on the fill, utilizing "dredge spoils" from the nearby Delaware River instead of the more traditional clay layer.[11] It was expected that these modifications would result in considerable savings.

Through a series of meetings and discussions, the PRPs' consultants convinced first the EPA engineering contractor and finally the technical staff at the EPA that their alternative solution to the Tybouts problem would provide as much protection from, and isolation of, the hazardous materials as more traditional solutions. This agreement was formalized in November 1985 in a document titled "Preliminary Agreement on Remedial Design." The RI/FS and ROD that were issued several months later embodied these concepts, although a failure specifically to authorize the use of dredge spoils in the cap was to complicate the final negotiations.[12]

THE END GAME. After completion of the RI/FS and ROD, technical progress at Tybouts slowed. Emphasis shifted to the federal district court in Wilmington. The year 1986 saw the most intensive legal activity of the entire Tybouts case. Discovery and pleadings among and between the various groups of defendants continued unabated. The year also saw an unsuccessful attempt by the federal government to amend its original complaint and add a group of third-party defendants to the four original defendants in the case.[13] This frenetic activity can be traced in the court's docket of the case, which records 858 separate entries for 1986, approximately 3.5 filings for each day the court was open during the year.

In the final stages of litigation the Tybouts case involved four categories of PRPs: the primary deep-pocket defendants (ICI, Stauffer, and New Castle County), Ward (who, while an original defendant, did not have deep pockets), the four "second-tier" defendants the government had unsuccessfully attempted to add to the original lawsuit, and a group of about twenty *de minimis* parties.

The case appeared to be moving toward closure throughout 1987. The government reached preliminary agreement in March with ICI, Stauffer, and New Castle County over their share of the remedial expenses. Working out an acceptable agreement with the third-party defendants (whose share in the remedy was approximately one-fifth of the total package) proved much more difficult, and delayed final resolution of the case another two years.

Until the government failed in its bid to add the second-tier PRPs to the primary lawsuit, Department of Justice attorneys handling the case had refused to negotiate directly with any of the third-party defendants. All contact with these corporations was filtered through the primary defendants. In mid-1987, this negotiating stance changed and Justice Department attorneys began to discuss settlement directly with the second-tier PRPs. The attorneys also opened negotiations with the *de minimis* parties, on the condition that they deal with the government as a group.

The negotiations took place over the summer of 1987, under somewhat unusual conditions. The government insisted that all negotiations be bilateral and confidential: each group of PRPs would cut a separate deal with the government, without any knowledge of the nature of the agreements being negotiated by other groups. The government had already reached agreement in principle with the primary defendants. By November 1987, the second-tier PRPs had reached preliminary agreement with the government on their percentage share of the total remedy and on most collateral issues. A group of *de minimis* parties negotiated consent decrees which were signed and delivered to the government in January 1988.

Despite the fact that all the major parties, and most of the minor parties, had reached closure with the government on the vast majority of issues by the end of 1987—including agreement on the basic design of the remedy and on the proportional shares that accounted for more than 90 percent of its cost—final consent decrees were not lodged in federal district court for another twenty-three months. The primary obstacles in the path of the final settlements during this period involved a number of relatively minor issues:

—Disagreements between the various groups of defendants concerning their respective responsibility for certain past costs,[14] and a few minor aspects of the future costs of the remedy;[15]

—Disagreements between the primary defendants and the government concerning the method of calculating stipulated penalties for nonperformance, and the nature of the monetary cap to be placed on oversight expenses incurred by the government that would be paid by the PRPs;

—A dispute between Ward (the landowner) and all the other parties concerning his right to compensation for use of his land adjacent to the dump during the lengthy remediation process.

Throughout the twenty-three months, several sets of consent decrees were signed and submitted to the government, only to be rejected because one group or another of the defendants refused to go along with provisions in the consent decrees of the others. The federal judge responsible for the case became increasingly frustrated with what he saw as intransigence on all sides and the minor differences between the parties standing in the way of a final settlement. The lawyers became progressively embittered with a process that had already consumed vast amounts of their energy and the resources of their clients.

The case was finally settled, probably because all participants were worn down by acrimony and by the energy the case had consumed over thirteen years, rather than because of any dramatic breakthrough at its conclusion. Indeed, given the narrow ground that eventually separated the parties, it would be difficult to label any final resolution as dramatic. One of the major bones of contention, the liability of the second-tier PRPs for past costs incurred by the primary defendants, was left for subsequent litigation.

Consent decrees were formally entered at the district court on December 21, 1988. There were a total of seven decrees, involving twenty defendants. The primary defendants—ICI, Stauffer, and New Castle County—agreed to pay for 79 percent of a cleanup estimated to cost about $20 million. They also agreed to construct the remedy.

Ward agreed to fund 3 percent of the cleanup, and to allow his land surrounding the dump site to be used during the remediation effort. Second-tier defendants du Pont and Budd agreed to contribute another 10 percent to the remedy, although they would not be involved in the actual construction of the remedy. More than a dozen of the *de minimis* parties signed consent decrees for fixed dollar amounts. The consent decrees in the aggregate accounted for approximately 95 percent of the predicted cost of remediation at Tybouts Corner.

Analysis

Despite the fact that the lawsuit in Tybouts predated any other activity in the case, scientific and engineering aspects of the cleanup dominated its early years. Region III's accommodation strategy made its major imprint on these early stages. About the time the RI/FS and ROD were being completed, primary control of the case moved from the engineers in Region III to the EPA's regional enforcement lawyers and to the Department of Justice in Washington. The strategy shifted to a prosecution approach, directly analogous to that applied in Region V in Laskin Poplar and Cliffs Dow.

The accommodation strategy is thus illustrated most clearly in the RI/FS and remedy-selection process. These procedures were open and participatory, a reflection of a Region III perspective toward Superfund cases that encourages negotiation and consensual resolution of disputes. Unlike in Region V, where technical data are not routinely shared with PRPs, and where there is a reluctance to incorporate PRP input into remedy selection, a conscious and substantial effort was made in the Tybouts case to keep the primary PRPs informed as findings emerged during the remedial investigation.

When the field activity for the RI/FS was about to be concluded, the assistant regional counsel in charge of the case wrote a letter to all the lawyers involved in the litigation, inviting them to participate in "consideration of cleanup alternatives for inclusion in the feasibility study."[16] PRPs were invited to submit written proposals for consideration. Subsequent working meetings between EPA and PRP lawyers and engineers were held to discuss the various proposals. One of the Region III RPMs who worked on Tybouts told us, "We're trying to develop something [in the RI/FS] that we have to go and sell to the PRPs, so they will go out and build it. You do that by keeping them involved in the decisions." Interviews with technical staff involved in the Tybouts case indicated a willingness to consider seriously the proposals of the PRPs.

Despite occasional lapses, relations between regional staff and the PRPs throughout the decisionmaking process were reported by all interviewees to be informal and nonconfrontational, as befitting the accommodation model. Strategic assumptions about the results of fostering this type of relationship were borne out—PRPs participated in a good-faith effort to conclude a mutually acceptable agreement. The result was preliminary agreement on the design of a complex, $20-million cleanup less than a year after the field investigation phase of the RI/FS was concluded. Our discussions with the PRP attorneys and with regional technical staff suggest that Region III's strategy of openness and accessibility contributed substantially to this result.

When the focus of activity moved from remedy selection in the regional office to negotiations over the consent decrees in federal district court, the situation changed appreciably. In court, most discussion took place between lawyers from the Lands Division of the Department of Justice and the PRPs; the strategy applied by the government was almost purely prosecutorial. The government's motivating assumption during these discussions was of PRP intransigence (an assumption that may have become something of a self-fulfilling prophecy) and the need to "stand tough" on all aspects of the discussions so that the PRPs would eventually capitulate. The moral imperative of joint and several liability dictated the necessity of total PRP assumption of the costs and risks of cleanup.

Two early decisions common in the prosecution strategy were responsible for much of the case history of Tybouts in its final years: the determination to "sue first and ask questions later," and the decision to proceed against three slam-dunk PRPs before conducting a more exhaustive PRP search that might have uncovered additional responsible waste generators.

There is little question that the decision to initiate legal proceedings at an early stage, with limited technical or evidentiary information on the site, influenced all that subsequently occurred. It should be remembered that the suit was filed in 1980, even before CERCLA became law. One early participant in that decision indicated that it was made as something of an experiment: "I guess the view was, 'Let's sue and see what happens.'"

Our interviews in Region III suggest that Tybouts was a chastening experience for all involved. A major lesson of the case, according to several EPA interviewees, is the error of commencing a lawsuit in Superfund cases until all other alternatives are exhausted. "Remember Tybouts" apparently is a potent response of technical staff in the region to those who have suggested recourse to a lawsuit in subsequent cases. From the perspective of the Superfund program staff, a lawsuit takes much of the staff's control over the progress of a Superfund cleanup and places it in a judge's hands. It also vests more power in the regional counsel's office and Washington attorneys. Administrative decisionmaking must advance according to the frequently slower pace of the court case, more sign-offs must be obtained to finalize agreements, and the use of administrative tools such as unilateral orders is greatly complicated. At a minimum, the presence of a powerful, generalist judge adds an element of unpredictability to all that follows.

The decision to sue just three PRPs in Tybouts was almost as consequential as the decision to initiate legal proceedings. Interviews with participants in that decision suggest that it was made in the absence of an adequate PRP search. Indeed, the only parties to the original lawsuit were the property owner (Ward), the dump operator (New Castle County), and one corporate waste-generator (Stauffer Chemical) that was inculpated by its own admissions in a congressional survey of hazardous-waste disposal practices. ICI was added to the original list of defendants after more investigation, reportedly at the insistence of the original PRPs.

As was shown in Region V, the practice of proceeding against a small number of PRPs characterized by deep pockets and clear liability has the predicted advantage of minimizing the government's administrative costs, at least in the short run. The government simply transfers the costs of the PRP search and accompanying investigation onto the defendants, who have an incentive (if not an imperative) to find other parties with which to share the costs of cleanup. However, these contributions will only come in the context of subsequent third-party lawsuits. The original defendants therefore must absorb not only the costs of their own defense, but also the expenses of the investigation and ultimate prosecution of the third- party actions.[17]

A perusal of the court docket suggests that the aggregate transaction costs of the third-party litigation in Tybouts were staggering. It is difficult to envision a more expensive or time-consuming investigatory process than the discovery conducted in this case.[18] Each one-line docket entry for a deposition (of which there were more than a hundred) implies thousands of dollars in legal fees, because the depositions were invariably attended by a dozen or more attorneys, and occasionally lasted several days.

Any complex case with as many evidentiary questions as Tybouts necessarily will incur substantial costs. But a government approach that included a careful PRP search, followed by uniform action against all those implicated, would surely have minimized some of the strategic, almost punitive, uses of discovery that characterized Tybouts. Removal of the artificial distinction between "ins" and "outs" would have eliminated some of the acrimonious and relatively petty disputes between the various groups of defendants that held up resolution for almost two years after the parties had agreed on all the major issues in the case.

The very early, pre-CERCLA decision to initiate legal proceedings in Tybouts may have been based more on happenstance and experimentation than on strategic choice. However, the desire to minimize government transaction costs probably influenced the decision to proceed against only three deep-pocket defendants in the case of a dump site that clearly was the responsibility of additional corporate waste-generators. This desire almost certainly accounts for the initial refusal of the government to negotiate with any but the original PRPs, and to restrict *de minimis* negotiations to those who would come forward and negotiate as a group.

A desire on the government's part to "have it all" with respect to minimizing both current expenditures from the Superfund and all future risk in the cleanup is critical to an understanding of the last years of the Tybouts case. Throughout the final negotiations, federal attorneys doggedly insisted that the risks inherent in a complex, ten-to-twenty-year cleanup be borne entirely by the PRPs, and they conducted negotiations in a manner which, while designed to minimize their own transaction costs, appears to have lengthened and complicated the process for all parties. For example, in their bilateral negotiations with the second-tier defendants, the Justice Department attorneys seemed either unaware of or uninterested in the fact that the "concessions" they were making in regard to indemnification and past costs would cost the government nothing. Instead, these costs would be paid by the primary defendants. This stance is consistent with a general position that the government's only interest in a Superfund case is the bottom line of aggregate PRP contribution, not its allocation. Indeed, the very conduct of bilateral, confidential negotiations was a strategy designed to maximize total dollar contributions by the PRPs by keeping each group in the dark about the deals being worked out by the others. There is no way of ascertaining if this strategy was successful in increasing the total percentage share the PRPs contributed to the Tybouts cleanup. It is clear, however, that it contributed to a longer, more contentious, and much more expensive negotiation process through 1987 and 1988.

The final disputes between the government and the primary defendants can be explained in similar terms. Two issues threatened to derail the entire negotiation process: the calculation of stipulated penalties in the event the major PRPs did not perform the cleanup as agreed in the consent decree, and the cap to be placed on the costs the government could charge the PRPs for its oversight of the cleanup. Given that ICI, Stauffer, and New Castle County had agreed to pay 79 percent of a cleanup predicted to cost upwards of $20 million, and had already expended hundreds of thousands of dollars in engineering and design work on that remedy, the position of the government in these final disputes seems at best to be unnecessarily parsimonious. The quibble was over a trivial amount, given the overall settlement already agreed to, yet the Justice Department position was unyielding.[19] This was certainly the perception of federal district judge Joseph Longobardi, who feared that this final intransigence on the government's part might force the case into trial. After hearing the government's position on these matters, he admonished the Department of Justice attorney: "Now it seems to me that you want your security blanket and you don't want to give up any parts of it. I mean, you want the whole hog. You have got to have some uneasiness about the results also. Otherwise, it wouldn't be negotiations, wouldn't be compromise. . . . You refuse to gamble, but you want everybody else to gamble.[20]

Part of the reluctance of the government to "gamble," even a little, can be explained by the bureaucratized nature of the legal machinery in Superfund cases. PRP attorneys expressed frustration that the cast of Justice Department lawyers handling the case changed frequently, so concessions worked out with one lawyer were forgotten or not understood by his or her successor, and none of the lawyers seemed willing or able to concede even the smallest item without obtaining multiple approvals from above.[21] This is perhaps unsurprising in light of an incentive structure for government attorneys—individually and collectively—in which the potential penalties for delaying or even unhinging a settlement are relatively low, while those accompanying the appearance of "giving away the store" are perceived to be very great.

Tybouts Corner, then, illustrates aspects of both the accommodation and the prosecution strategies. Most of the major issues in the case were resolved within eighteen months of the conclusion of the exceedingly tardy RI/FS. The major PRPs had agreed to conduct the cleanup; the EPA had agreed with the basic outlines of the remedy proposed by the PRPs; the various parties at the site had agreed to pay for more than 90 percent of the cost of remediation and other parties were available to absorb additional costs. These decisions were reached through relatively cordial negotiations with the PRPs that were conducted in large part by Region III personnel.

After transfer of the activity to the Justice Department and recourse to a prosecution model of Superfund implementation, three more years of acrimony passed and substantial additional transaction costs were run up before the case finally was resolved. As in the Region V cases, the government ultimately compromised at the end of the process. The 100-percent solution was not obtained, and the government assumed some of the indemnification costs that were of such concern to the primary defendants. Given all that had already been agreed to in 1986, we suspect it would be difficult to justify the additional delays and costs in terms of whatever value added the government received in the final consent decrees of 1989.

Harvey and Knott Drum

The Harvey and Knott Drum Site is located in a predominantly rural section of New Castle County, just 150 yards from the Maryland-Delaware line and not far from the Tybouts Corner Superfund site. Across the road and to the north of the dump is Shelly Farms, a residential development of about one hundred homes; to the south and west are wetlands and swamps. Harvey and Knott Trucking, Inc., owner of the site, operated it as an open dump from 1963 until its closing in 1969. The dump received sanitary, municipal, and industrial wastes such as packing material, trash, and canteen waste; it also received sludge, paint pigments, and solvents, which were stored in drums, burned, or poured onto the ground or into trenches.

The dump site sat, unused and virtually unnoticed, from its closing in 1969 until January 1981, when it was photographed during an overflight by the Maryland Waste Management Agency. The photographs revealed dying vegetation and what appeared to be barrels and drums stacked at the site. After ascertaining that the dump was located in Delaware, the Maryland agency informed Delaware officials and the EPA.

The process leading up to agreement between the major PRP at the site—General Motors—and the EPA on the Harvey and Knott cleanup is in almost complete contrast to the events that characterized the later stages of the Tybouts Corner case. From an engineering perspective, the cleanup of the Harvey and Knott site was expected to be smaller, simpler, and much less costly than the Tybouts cleanup. Public concern and media attention were less intense. Rather than the several dozen PRPs present at Tybouts, the government dealt with a handful at Harvey and Knott; in the end, all serious decisions were worked out between the EPA's Region III and General Motors. Perhaps partly as a result of these characteristics, the negotiations between the government and General Motors were uniformly described as smooth and productive. The overall atmosphere was characterized in our interviews as one of mutual trust and even goodwill. This case could serve as a defining example of the accommodation strategy.

Summary of Key Events

A joint federal-state team inspected the Harvey and Knott dump in November 1981, and recommended sampling soil at the site, as well as groundwater and water from private wells at the nearby Shelly Farms subdivision. Tests were conducted in March 1982, and the EPA received an initial analysis in mid-May. The first set of tests indicated elevated levels of lead and polychlorinated biphenyls (PCBs) in five of the residential wells. However, subsequent tests failed to confirm these findings.

After an unsuccessful attempt to involve the Harvey and Knott families (owners of the site and the trucking firm that brought waste to it) in the endeavor, the EPA began a removal action in which it installed a security fence, sealed and covered the leaking drums, and started an extensive sampling program. In addition to taking additional samples of soil at the site and water from nearby residential wells, the EPA drilled six on-site monitoring wells. The removal was completed in late August 1982, at a cost of $55,000.

Autumn 1982 saw the EPA involved in an investigation to determine the responsible parties at the site. In addition to various Harvey and Knott family interests, General Motors and Chrysler received section 104(e) notification letters, as did several companies whose names were found on barrels at the site.[22] Harvey and Knott had maintained almost no records of the source, nature, or volume of waste it had dumped. As at Tybouts Corner, the evidence the EPA ultimately relied upon consisted of testimonial reports from truck drivers working for the site owners, and from the dump manager. These investigations led the EPA to limit the PRPs to the various individuals and businesses associated with the Harvey and Knott families, and two deep-pocket waste generators, General Motors and Chrysler.[23]

In early 1983, the EPA sent notices to the PRPs it had identified in its investigation, indicating that the agency intended to commence work on the RI/FS, and inviting their participation. Only one member of the Harvey family indicated an interest in negotiating; the rest of the PRPs, including General Motors and Chrysler, demurred. Region III obtained authorization to begin an RI/FS in April 1983, and work began in June. In addition to the investigatory and engineering work contemplated in the RI/FS, the EPA did additional construction at the site: fencing was installed, erosion control measures were taken, surface barrels containing waste were consolidated and transported to a RCRA disposal site, and empty barrels were taken to a recycling facility. A contractor was retained by the EPA to conduct the RI/FS. In September 1983, Harvey and Knott Drum was added to the NPL as number 696.

About the same time Harvey and Knott was added to the NPL, General Motors retained a new attorney to represent it in the case, and hired an environmental engineering consulting firm to do a "shadow RI/FS" at the Harvey and Knott site. General Motors' attorney notified the EPA's project manager at the site that the company wanted to negotiate a voluntary cleanup. Through 1984 and the spring of 1985, the technical data generated in the government's remedial investigation was shared with General Motors, whose lawyers and technical consultants were also present at various meetings at which alternative methods of site cleanup were discussed. By late spring 1985, a "Preliminary Endangerment Assessment and Remedial Alternatives" document was given to all the PRPs for comment.

Unsurprisingly, given its continuing participation in the formulation of the RI/FS, General Motors' reaction to the draft document was favorable. In its formal comments, General Motors made only minor suggestions about the proposed remedial alternatives. This same pattern of interchange between the EPA and General Motors was evident when the government's contractor submitted the draft RI/FS in August. General Motors' comments were broadly positive, with some specific suggestions concerning the groundwater remedial action and several other matters.

Chrysler, on the other hand, decided (in the words of one interviewee) to "stonewall." Its representatives did not participate in the technical discussions carried on throughout the RI/FS period. While Chrysler did comment on the draft RI/FS, those comments were provided to the EPA after the deadline for submission, only five days before the final RI/FS and ROD were issued. In these comments, Chrysler indicated almost total disagreement with the RI/FS, suggesting that it was "fundamentally flawed" because no link was established between groundwater contamination and "potential receptors."[24]

The ROD, signed on September 30, 1985, provided for a remedy that included cleaning of the onsite drainage pond by collection and treatment of the surface water; removal and off-site disposal of contaminated sediments, sludges, and bulk wastes; removal and off-site disposal of all crushed or intact surface drums and other debris and sludges; installation of an extraction and treatment facility to collect and remove contaminants in shallow groundwater, and then use of the treated groundwater to flush contaminants from soils in the site.

RODs are not noted for being expecially explicit as to the remedy decided upon, but this one was noteworthy for its vagueness on how subsequent groundwater remediation was to be achieved:

EPA is not prepared at this time to determine the appropriate level of groundwater corrective action at this site. Operation of the extraction/treatment/reapplication facilities will [continue] for an estimated five years and should substantially reduce the amount of contaminants in the ground water in the vicinity of the fenced area and reduce the plume.[25]

An EPA official involved in the discussions said that the ROD was "not terribly detailed because it wasn't all that clear what was there. Rather, there was a 'concept.'" The "concept" was to pump and treat the groundwater and then, after several years, determine what (if anything) needed to be done to clean up the groundwater more permanently.

POST-ROD ACTIVITIES. By the time the record of decision was signed, it was reasonably clear that General Motors would build the selected remedy, that Chrysler would not participate, that a mixed-funding arrangement would supplement General Motors' contribution, and that the EPA would proceed against Chrysler in a subsequent cost recovery action to obtain reimbursement for the monies drawn from the Superfund. The specifics of the mixed-funding deal, particularly the percentage share to be borne by General Motors and by the Superfund, remained to be established, as did the specifics of the remedial design.

As expected, Chrysler refused to participate in the negotiations that followed issuance of the ROD, arguing that the company had no liability because it had not generated any of the waste deposited at the site. The Harvey and Knott interests did not participate in this process either, although this was related more to their financial limitations than to any denial of liability. The bilateral General Motors-EPA negotiations were conducted primarily between General Motors' lawyer and the assistant regional counsel assigned to the case, along with the site's remedial project manager. According to all reports, these negotiations went "very smoothly." The concept of a mixed-funding arrangement was quickly agreed upon, although the logistics of drafting the required documents and obtaining the necessary authorizations in this untried process consumed much administrative time.

The proportional share that General Motors ultimately negotiated—67 percent of cleanup costs—was lower than the then-current headquarters guidelines for mixed funding, which suggested a minimum 80 percent PRP share. It was substantially lower than the 100-

percent PRP contribution subsequently demanded by the EPA in similar cases.[26] According to participants in the negotiations, the figure represented a rough estimate of General Motors' volumetric share of the waste at the site, plus a "premium."

The other major subject of negotiations between General Motors and the EPA during the fall of 1985 was design of the remedy. As indicated previously, the ROD was not specific about the later stages of cleanup at the Harvey and Knott site; this lack of specificity was most evident in the discussion of groundwater remediation. But even the initial soil and drum removal and the "pump-and-treat" system were not set out in detail. According to a subsequent document, "The chosen remedy is complex and will go beyond the specifications set out in the Record of Decision as further information becomes available during the initial phases of implementation."[27]

REMEDIAL DESIGN. Throughout October and November of 1985, General Motors' technical consultant worked on a draft work plan for the cleanup, in consultation with the company's technical staff. The chief of environmental affairs at General Motors took the lead on behalf of the company in this process, and was responsible for much of the broad design of the remedy. Like the ROD, the work plan embodied a concept, the details of which were to be worked out as more and more information became available to inform the engineering analysis. The amount of soil to be removed was not indicated, and the specifications for soil and groundwater quality were left unresolved, as was final design of the middle and final stages of remediation. The decision on whether or not an impermeable cap would be necessary was deferred until the impact of the pump-and-treat procedures could be assessed.

The draft work plan finally submitted to the EPA in mid-December 1985 was labeled "part of the ongoing settlement negotiations." A draft consent decree went to the agency in January 1986. In early April, the EPA wrote to General Motors, informing the company of the agency's judgment that "overall" the work plan was a "well written, technically sound document." The letter conveyed only one major concern: how decisions about the ultimate design of the groundwater recovery system would be made.[28] Beyond the level of specificity in the work plan, and the proportional share of cleanup costs to be borne by General Motors and Superfund, the only other major issue in the negotiations concerned whether all ARARs (applicable or relevant and appropriate state and federal laws and regulations) needed to be met as part of the cleanup.[29]

The EPA notified General Motors in late April that it had approved of the draft work plan. Both parties executed a written agreement in principle concerning cleanup of the Harvey and Knott site on May 5; details were finalized over the summer. The total cost of the remedy was estimated at $10 million, although provision was made for subsequent requests for CERCLA funds if the expenses exceeded that estimate. The consent decree was formally approved by the federal district court in Wilmington on December 18, 1987. The Department of Justice subsequently filed a section 107 cost recovery action against Chrysler for $3.3 million in past costs (the amount of Superfund monies authorized for the site) and one-third of future costs involved in cleaning up the site.

Analysis

Harvey and Knott illustrates the flexible and open approach to Superfund negotiations that is a hallmark of the accommodation strategy. As in Tybouts, the technical data generated throughout the RI/FS was made available to the PRPs in a timely fashion. General Motors, the only PRP that showed interest in cooperating on the remedy, had input into the process of fashioning that remedy. General Motors' lawyer and the EPA assistant regional counsel assigned to the case played essential roles in the design of the consent decree and the apportionment of shares between General Motors and Superfund. But the critical technical issues were left largely in the hands of the engineers from General Motors and the EPA, and their respective consultants. All indications are that these discussions went forward on a cooperative, rather than adversarial, basis.

The accommodation strategy is based upon a decisionmaking process that is informal and open to PRPs. Applied to a series of discrete issues in Superfund cases, this process builds confidence on both sides, and contributes to the success of subsequent negotiations over potentially more contentious matters. This approach produced the intended result in Harvey and Knott: the degree of mutual trust between the EPA and General Motors generated by this process is best illustrated by the open-ended nature of the remedy specified in both the ROD and the work plan attached to the consent decree. As indicated above, only the broad concept for the remedy was set out in these documents. Final design of the groundwater system, by far the most expensive element of the remedy, was left for determination after sufficient experience and information could be gleaned from earlier steps in the process. The final cleanup standards were also left to future determination.

Unlike PRPs in similar negotiations, General Motors seemed willing to accept this level of uncertainty relative to the total possible cost of the remedy it was agreeing to build. Indeed, several interviewees suggested that General Motors was signing a "blank check," the amount of which would only become clear several years into the cleanup. Of course, the one-third share of the remedy to be paid from the Superfund provided a disincentive for the agency to add unnecessarily costly elements to the final remedy. General Motors thus had some reason to believe that the agency would remain reasonable in the years ahead. Still, there is little question that the company was gambling on the continuation of what had been a good working relationship between its engineers and those of the agency.

The EPA, for its part, was also taking a risk since the lack of specificity in the ROD and (especially) the work plan, together with the yet-to-be-determined cleanup standards, left the agency open to the charge of granting too much discretion to the PRPs, and thus inadequately protecting the public's interest in a full cleanup. In order to take such risks, the remedial project manager and assistant regional counsel assigned to the case needed to be assured of the support of their superiors, and the administrators at the regional level had to be reasonably sure of their political position within the agency. Also, the officials at EPA headquarters who ultimately approved the deal needed the self-confidence that results from some minimal insulation from the political winds that blow from Capitol Hill and the environmental community. Our observations in other regions, and discussions with members of the national Superfund community, suggest that the confidence necessary to take such risks in the conduct of negotiations and in the design of PRP-led cleanups is often in short supply at all levels of the Superfund program.

Sharing of costs and risks of cleanups, primarily through use of mixed funding and the other statutory devices that encourage settlement, is another defining characteristic of the accommodation strategy. The case of Harvey and Knott presented an ideal context for this type of risk- and cost-sharing. A consent decree specifying a PRP-led cleanup utilizing mixed funding was attractive to all concerned. It allowed General Motors to get on with the cleanup without having to bear its total cost, without incurring the transaction costs inherent in litigation, and with some control over the design and construction of the remedy, and it promised the regional office prompt initiation of the cleanup, without the transaction costs and delays inherent in litigation. Chrysler's refusal to negotiate or contribute to the cleanup raised the strong possibility that the government's share of the cleanup could be obtained in a subsequent cost recovery action. One Region III official said, "We saw mixed funding as just another approach to get a settlement in these cases. We didn't see any reason in standing on principle and threatening court action, especially after *Tybouts*. We are proud of the mixed-funding agreements during this period."

Interestingly, we heard in Region III none of the common complaints about the endless bureaucratic hassles involved in mixed-funding settlements. Harvey and Knott was one of the very first mixed-funding agreements, and thus required a substantial amount of legal and administrative work of a technical nature. But, as in Tybouts, the basic structure of the mixed-funding settlement seemed to cause little difficulty at any level.

Harvey and Knott Drum involved a small number of PRPs—only two corporate deep-pocket parties, plus the Harvey and Knott interests. There is little question that the configuration of PRPs at the site made it ideally suited for a mixed-funding settlement. In terms of size, seriousness, and the potential costs of cleanup, Harvey and Knott fell into the middle range of NPL sites. Particularly given the small number of PRPs, the estimated $10 million cost of the cleanup puts it well beyond the type of case exemplified by Davis Farm in Region IV (see chapter 5), in which the case was settled for its nuisance value and disputed legal issues were dropped by PRPs in order to avoid transaction costs. The Harvey and Knott case can be more usefully compared to Cliffs Dow in Region V (see chapter 3), where similarly cooperative PRPs were treated like bad guys who could not be trusted.

Assessing the Accommodation Strategy

The accommodation strategy calls for the government to treat PRPs with informality, fairness, and a modicum of good will: PRPs receive and are able to comment upon technical studies; decisionmaking at Superfund sites includes a process of negotiation rather than the presentation of nonnegotiable demands; the government shows a willingness to induce settlement by assuming some of the costs and risks of cleanup. The assumption is that such a process will lead to cooperative relationships through which resolution of the many scientific and economic issues at Superfund sites can be achieved with greater speed, and with fewer transaction costs than would be the case with a more coercive, adversarial approach. To what extent did these assumptions prove correct in our case studies?

The scientific and technical questions involved in remedy selection at both Tybouts Corner and Harvey and Knott Drum were resolved through processes that closely approximated those called for in the accommodation strategy. In both cases, PRPs were given raw technical data as it was generated, drafts of reports, and other administrative documents. In both cases, the remedies were developed through a process that included both PRP and EPA scientists and program staff. EPA project managers listened to PRPs and modified cleanup requirements in light of concerns and questions raised. This is not to say that the regional staff was somehow "captured" by the business interests at these sites; on many issues the agency disagreed with the PRP position, and acted accordingly. But remedial project managers and regional lawyers kept the PRPs informed, taking their concerns and proposals into consideration when making decisions.

The accommodation strategy had the intended effect. To a substantial degree, PRPs responded to the invitation to participate by presenting reasonable proposals and engaging in good-faith negotiations. Remedy selection and other scientific and technical issues were resolved speedily, with relatively low transaction costs and minimal acrimony. This informal process of negotiated decisionmaking was applied to financial questions, as well as scientific questions, in Harvey and Knott Drum. As a result of the mutual trust established through the remedy selection process, General Motors and the EPA agreed to a multimillion dollar settlement in which the costs and risks to both parties were substantial.

In the case of Tybouts, the accommodation strategy was not applied beyond the technical issues resolved at the regional level. Final, formal resolution of financial issues was left largely to attorneys in the Lands Division of the Department of Justice in Washington. Negotiations with PRPs were characterized by reliance on joint and several liability to achieve total government victory, followed by nonnegotiable demands, coercive threats, and legal maneuvers. As in the two case studies from Region V, this approach led PRPs to dig in their heels. They initiated a flood of third-party litigation and engaged in a variety of tactics that slowed the process of dispute resolution to a crawl while adding immeasurably to the transaction costs to all parties.

We had hoped to find a large, complex site in Region III where the accommodation strategy was applied from start to finish. While Harvey and Knott Drum is hardly an insignificant site—it will ultimately involve expenditure of more than $10 million—it is not a case of the magnitude of Tybouts or Laskin Poplar. We are left, therefore, with the question of how a strategy of accommodation would play out were the stakes substantially higher. We suspect that this strategy is seldom applied in *any* region to such "hard" cases. Our failure to find a large, complex case in which accommodation was used throughout may not have been accidental.

Region III had substantial control over events in the early engineering studies and during remedy selection, and was thereby able to employ its preferred accommodation strategy. But the final resolution of big, expensive, and highly political cases is almost always by means of a consent decree in federal district court, a document that is inevitably negotiated by attorneys in the Lands Division of the Department of Justice. Our own interviews at the Justice Department support the perceptions of the department we obtained from Superfund regulars around the country: it is committed to a prosecution strategy in Superfund, with the accompanying decision rules, reliance on joint and several liability, and assumptions about PRP behavior. Unless regional offices are given more authority in negotiating settlements in big Superfund cases—thereby reducing the unidirectional influence of the Justice Department—it seems likely that an accommodation strategy will be used infrequently, if at all, in the final resolution of such cases.

Five

Public Works

WHEN ASKED about the character of hazardous waste cleanup in Region IV, many members of the Superfund community express a similar view: in Region IV the engineers are dominant, and the concern is to get the shovels moving and the toxics off the ground.[1] The mandate is to clean up the site now and let the lawyers and accountants worry about money and legalities later. Region IV reportedly employs the emergency removal process in circumstances in which the remedial mechanism would be used elsewhere, and publicly funded cleanups are common. Politically, because it is located in the Deep South, an area that tends to be conservative on environmental matters, the office is also reputed to be reluctant to use Superfund's enforcement powers against corporations or government entities.

Regional reputations are based on past, rather than current practices. When we visited Region IV, we found little dispute as to its historical reliance on publicly funded removals. But regional officials also reported changes in personnel and policy that may have led the region away from the public works approach to Superfund implementation. At the same time, the region does have a distinctive character: those who deal with the Environmental Protection Agency in Illinois and then cross the river into Kentucky, for instance, report that working with Region IV is very different from interacting with Region V. Of more relevance to our research, the region contains a number of

The field research for this chapter was conducted by Phillip Cooper, who also wrote its first draft.

sites where a public works strategy was used, and in which we can examine its operation.

Regional Characteristics and Strategic Choices

With headquarters in Atlanta, Region IV includes most of the states of the Deep South. The region contains about the same number of National Priorities List sites as Region III—13 percent of all NPL sites. This makes its Superfund program large compared to that of most other regions, but small compared to that of Region V. Staff turnover has been a significant problem in the region, particularly among Superfund project managers. However, the Office of Regional Counsel (ORC) has experienced a relatively high level of staff stability. Whether as the result of a more manageable mission or the Southern traditions of informality and gentility, the regional office does not give one the impression of a large, impersonal bureaucracy. Attorneys for potentially responsible parties (PRPs) describe Region IV personnel as less hostile and aggressive than, for example, those in Region V.[2]

Region IV's national reputation is based on an engineering approach that de-emphasizes the liability aspects of the statutory scheme in favor of using trust fund resources to achieve rapid physical improvement "on the ground." One regional official we interviewed summed up this aspect of regional strategy with the slogan, "We move!" The statutory tool of choice is the emergency removal procedure, in which Superfund trust monies are applied to speedy, public works-style cleanups. After cleanup is accomplished, regional lawyers may then seek compensation from the responsible parties.

The typical removal involves what its name implies: waste is removed from a site and taken to an approved hazardous waste disposal facility. This process allows sites to be cleaned up without the delays and legal maneuvering inherent in the more ponderous remedial approach to cleanup. But, at least initially, this process usually involves the expenditure of federal, rather than private, funds.[3] Historically, Region IV has not been aggressive in obtaining private-party participation in the conduct or financing of cleanups. Indeed, the region has been criticized for its failure to proceed against PRPs in Superfund cases, and consequently has intensified efforts to recover costs from PRPs after removals are undertaken.

Our interviewees pointed to several factors that have shaped Region IV's distinctive orientation toward Superfund. As in most complex organizations, tradition undoubtedly played a part.[4] The nature of dump sites in the region may make them especially good candidates for removal procedures; several interviewees indicated that there is a preponderance of small sites with nonviable PRPs.[5] The concentration of approved hazardous waste-disposal facilities in the South makes removals less expensive than in other regions of the United States.[6] However, we believe that two other factors are also central to the adoption of a public works strategy in Region IV: the professional orientation of the office, and the nature of the states with which that office must deal.

Unlike in Region V, where lawyers predominate, program staff control Superfund in Region IV. The technical staff is dominated by engineers who place primary emphasis on cooperative problem-solving and action—an orientation that is frequently at odds with legalistic concerns. The Office of Regional Council in Region IV plays a similar role to that of the ORC in Region III: it advises on legal matters but has little influence on site-specific decisions prior to efforts at cost recovery. As in Region III, ORC attorneys describe their relationship to program managers as one of lawyer and client; program staff set the priorities and make the management decisions. We noted among project managers an impatience with legal procedures and time-consuming negotiations, and an accompanying desire to get the bulldozers moving.

We believe that selection of the public works approach to Superfund implementation in Region IV has also been driven by the degree of commitment of states in the region to hazardous waste cleanup, and the amount of resources they can allocate to the problem. The influence of state environmental officials varies from state to state, however. Florida, with substantial worries over groundwater, is the most environmentally conscious state in the region. Some interactions between Florida and Region IV reportedly are congenial; others are not. Relations between the state of Georgia and the EPA regional office are seen as particularly difficult. Other states in the region have so few resources to devote to environmental problems that they demonstrate little guidance, leadership, or even interest. Because of the comparatively low level of environmental concern among southern states, Region IV enjoys a somewhat freer hand in Superfund cases than do regions faced by aggressive state environmental agencies protective of their own organizational prerogatives and stringent state standards.

The lack of state resources available for cleanup is also a significant factor in Region IV. Most southeastern states, in the words of one EPA official, "have no money. We have money; they don't." While the availability of federal resources can make recourse to a public works approach attractive, this is not always the case. At times, the lack of state funds can work against use of a public works strategy, since Superfund legislation requires that a 10-percent "match" be made by the state in some circumstances when cleanups are financed with federal Superfund monies. Remedies must be shelved if state funds cannot be found to finance the match.

Within this context of constraints and opportunities, Region IV developed a distinctive approach to the implementation of Superfund. This strategy posits the primary goal of the Superfund program as the prompt cleaning of hazardous waste sites. The statutes establishing the Superfund program include just the tool for such immediate action: the emergency removal powers.

Removals, at least as practiced in Region IV, typically involve an EPA-only operation in which regional staff decide what needs to be done at a site and then hire contractors to do it. Intended to deal with imminent hazards rather than remediation of sites that pose no immediate threat, the removal device allows the EPA to initiate timely physical improvements at hazardous waste sites. Region IV's implementation of Superfund has focused on an aggressive use of removal powers at these sites, some of which look much like typical removal sites in other regions, others which would almost certainly be handled as remedial actions elsewhere.

The emphasis on government action in the public works strategy frees the EPA from the major impediment to speedy site remediation that dogs other implementation strategies: the necessity of dealing with PRPs. Once a proposed cleanup is classified as a removal, regional staff can design and implement the remedy. There is no need to seek agreement from private parties, authorizations from lawyers in EPA headquarters or the Department of Justice, or consent from judges.[7] While the subsequent cleanup can proceed expeditiously, attempts to recover costs from PRPs can be slow and costly.

Use of removals is limited by statute and by a fiscal and political environment shaped by the finite pool of resources for government-financed cleanups. The authorizing legislation in Superfund restricts the emergency removal authority to sites posing an "immediate and substantial danger to public health or the environment."[8] Also, it places a limit of $2 million on Superfund expenditures on any one removal. These impediments to the use of removals have been by-passed in the region through use of a novel, unauthorized device. A combination of the removal and the remedial approaches to cleaning up waste sites, it is termed (in Region IV and elsewhere) the *removial*. A removial is a fund-financed cleanup—employing removal authority and removal contractors—that is undertaken at a site that is too large to be completed within the $2 million maximum, or that does not meet the endangerment requirements of the statute, or both. Use of removials sometimes requires regional officials to engage in conscious overestimation of the risks posed by a site so that these risks can be portrayed as constituting an "immediate" hazard. The removial option may also involve intentional underestimation of the cost of a cleanup when the predicted expenses do not greatly exceed $2 million or, when costs are predicted to be higher, the segmentation of one removal into several, each of which is estimated to cost less than $2 million.

Removials are controversial. Advocates contend that they can save substantial time and money at a site. If the sites are small enough and the cleanup is fast enough, removials result in unambiguous (and comparatively small) bills for PRPs, rather than contentious negotiations over unknown and potentially unlimited cleanup costs. However, a 1988 report by the EPA inspector general rejected the region's claim to having found an innovative approach to cleanup, and instead found that it had improperly used fund-based emergency removal procedures when it could have stabilized sites and proceeded with a proper remediation conducted by the state or a viable PRP.[9]

The cost recovery unit in Region IV serves as a vehicle for reconciling the region's long-standing removal mentality with the need for fund reimbursement and national pressure to adopt a more enforcement-oriented approach. The cost recovery unit determines the costs of cleanup, identifies the responsible parties, and assures the provision of sufficient financial information to support any administrative orders and cost recovery litigation. Attorneys are cast in the role of bill collectors who ensure PRP accountability, while program staff on the technical side continue to pursue the goal of "getting the stuff off the ground."

The sites we examined in Region IV exemplify this approach. Lee's Lane Landfill in Kentucky was one of the first Superfund sites in the region. This NPL site, while not as formidable as Laskin Poplar or Tybouts Corner, still provides a representative picture of the use of the public works strategy at a major site involving multiple PRPs. Because of statutory and financial constraints on the expenditure of Superfund monies, it may represent the outer limit of the applicability of the public works approach.

The simple site, Davis Farm, in Georgia, involved removal of material improperly transported from a nearby Resource Conservation and Recovery Act (RCRA)-licensed disposal facility. It was not a large site in dollar terms; indeed, it was not even an NPL site. Nevertheless, we chose to examine Davis Farm because it was representative of the class of site that is perhaps best suited to a public works approach. In many of these smaller cases, the removal eliminates the need for further remedial action, and the site is never listed on the NPL. Without an examination of processes at these smaller sites, we would be unable to assess the public works strategy in the context of cases to which it is arguably most applicable.[10]

Lee's Lane Landfill

Lee's Lane is a 125-acre dump site on the banks of the Ohio River, just south of Louisville, Kentucky. It had served since the 1940s as a repository for various forms of industrial waste. Residents of the adjoining Riverside Gardens neighborhood had been complaining to state officials about the landfill since the mid–1960s. No serious response was made until migrating methane gas from the landfill started causing small explosions in basement water heaters. Kentucky environmental officials closed the landfill in 1975 and were forced to relocate some of the families from Riverside Gardens. Later, discovery of some four hundred exposed chemical drums in the floodplain of the Ohio River, and recognition that the landfill contained a large quantity of waste that could cause heavy-metal contamination of the river and injury to local residents, led to further action.

In many ways, the EPA's actions at the Lee's Lane Landfill represents a pure case of the public works strategy; Lee's Lane is also an example of the removial approach that brought Region IV under administrative and congressional scrutiny. A PRP-financed cleanup of the site was proceeding during the mid–1980s, with a PRP steering committee negotiating with the EPA over conduct of a remedy specified in a 1986 record of decision. Unhappy with the pace of the negotiations (which had only been under way six months), and anxious to show some results "on the ground," Region IV notified the PRPs that the site would be cleaned up by removal contractors, using the emergency removal provisions of the Comprehensive Environmental Response, Compensation, and Liability Act (CERCLA). The $3-million cleanup was concluded within thirteen months of the completion of the ROD, in late 1987.[11] However, the cost recovery efforts of the agency were still pending in early 1991.

Summary of Key Events

In December 1979, Region IV prepared a potential hazardous waste site identification and preliminary assessment of the Lee's Lane site. A potential hazard was found in the form of groundwater contamination. In an assessment the following January, flammable wastes were found at the site. Region IV and the State of Kentucky worked together to monitor the floodplain during early 1980, contemplating remedial action and keeping alert to the possibility of floods.

In April 1980, Kentucky environmental authorities ordered the site owners to remove the drums at the Lee's Lane dump. State officials inspected the site again in June 1980, reported that about four hundred drums continued to be present in the river floodplain. They also noted the presence of corrosive and toxic sludge and indicated that migration of flammable gasses was occurring at the site. A gas-venting and incineration system was installed later in 1980 to alleviate this problem. Unfortunately, the incineration portion of the system never functioned effectively and the equipment was not maintained.

An EPA field investigation team inspected the site in February 1981. Its preliminary report called for removal of the remaining exposed drums, posting of warning signs, and further investigation of groundwater problems. Lee's Lane was included in Region IV's list of candidates for the first round of emergency actions under CERCLA. The site was placed on the NPL (as number 260) in December 1982.

Formal Superfund actions came the following year. A remedial action master plan was completed in May 1983. In June 1984, the EPA sent notices to several hundred PRPs. A remedial investigation and feasibility study was finalized in April 1986, without involvement of the PRPs. In the RI/FS, the agency found no immediate health risk from the landfill, although there was found to be potential heavy-metal contamination of groundwater and further gas migration from the dump site. Several of the more sanguine conclusions of the remedial investigation/feasibility study were challenged by residents of the community, but these concerns did not appear to affect remedy selection.[12]

The September 1986 ROD called for a remedial alternative that included repair of the gas collection and incineration system, evaluation of alternate water-supply needs, cleanup of surface pollution, capping of hot spots, riverbank protection controls, installation of gates and warning signs, and long-term operations and maintenance (O&M) activities. The dump was not to receive an impermeable cap. The remedy selected was estimated to cost $2.9 million, while the next step up in the RI/FS list of alternatives would have entailed an expenditure of $42.6 million.[13] The choice of remedy received serious criticism at the time only from the Riverside Garden residents.[14]

FROM ROD TO CLEANUP. The PRPs, as noted above, did not participate in the RI/FS. However, a steering committee had been formed soon after notice letters were sent, and it appeared likely that the remedy chosen in the ROD would be constructed by the PRPs. While the original notice process identified several hundred parties, the locus of activity ultimately involved about two dozen. Representatives of these parties met several times at the beginning of the process; eight coalesced into a steering committee chaired by a representative of B.F. Goodrich.

Negotiations moved slowly in the first months after the ROD was issued. The steering committee had difficulties organizing the finances. Although the total amount at issue was less than $3 million, there was not enough support from the rest of the group of PRPs for the eight to move forward. In the words of one participant, the most active PRPs "were not willing to put up the money and [then] sue the others."

Although it seemed likely that the ROD ultimately would be implemented by the PRPs, Region IV abruptly changed course in early 1987 and decided to do the cleanup itself. The cleanup was conducted by removal contractors between March and October of 1987. The entire cleanup was concluded just thirteen months after the ROD was signed. This action was done in the face of a formal finding of no imminent danger in the ROD, and in the context of an estimated total cost that exceeded the $2-million ceiling by almost 50 percent.[15]

Just a year after the EPA's selection of a remedy for the site, the PRPs found themselves discussing an allocation of the agency's already-expended costs, rather than attempting to divide the potential expenses of a remedy they would subsequently implement. Our interviews with PRP attorneys suggest that this new situation was in some ways advantageous to the PRPs. The EPA had simply done the work and sent them a bill. Since the total cost of the cleanup was less than even the original $3-million estimate, most PRPs were content to pay. Also, the negotiation process had been simplified. One of the PRP attorneys said, "After the cleanup, the bill was clear, unlike the situation before EPA took over. There was a relatively fixed dollar amount and now we had 'judgment day.' People who had been thinking, 'Let me sit on my cash for now and I'll pay from next year's budget,' saw the need to settle."[16]

The PRPs were well along in the allocation process when the government indicated that additional PRPs needed to be considered. The original PRPs asked for more information about these new parties but got more than they bargained for. The government produced a complete allocation of costs among the PRPs, based upon tiers of responsibility: 10 percent for the four largest contributors of waste at the site, and shares ranging from 2 to 8 percent for the rest.

The EPA allocation complicated the ongoing negotiations among the PRPs and, according to one participating lawyer, undermined allocation discussions: "Some companies that had offered more money immediately insisted on no more than what the government estimated." The larger companies were unhappy because they were being asked to pay a bigger share than had already been agreed to. The government turned its assessment into a nonnegotiable offer, however. Ultimately, the companies accepted the proposal because, as one attorney told us, it was "not worth fighting over that kind of money."

FINAL SETTLEMENT. Negotiations on the consent decree were more or less concluded about a year after construction was completed, by December 1988. The agreement returned $2.5 million of a remedy which, with administrative costs, had cost the government $3.2 million. The balance was to be recovered from nonsettlers. The EPA continued to add new parties well into 1989. The government did not file the consent decree until October 1990.

PRP attorneys whom we interviewed suspected that two difficulties delayed finalization of the agreement. The first was operations and maintenance costs, an issue that was still unresolved in early 1990.[17] The second difficulty involved the concurrent investigations of contracting practices in the region, particularly the use of removal contractors in remedial actions. The PRP attorneys contended that the ongoing investigation of regional removal practices caused Region IV to be cautious about pressing toward closure at sites where the investigations might upset or otherwise adversely affect the agreements.[18]

At the same time the agency filed the consent decree in court, it added the Ford Motor Company and Dow Corning as PRPs in the case, reportedly as a result of information developed in connection with the EPA's investigation of another local site. The EPA offered the two new parties an opportunity to join the consent decree for a 10-percent share of the costs. Despite the fact that both Dow and Ford were relatively small contributors to the Lee's Lane dump, and neither was initially regarded as a recalcitrant, the government contended that the two should pay the same share as the biggest waste-generators because they were "late joiners." Furthermore, interest was to be assessed against Ford on the theory that it had been in the case earlier and should have been aware of its potential liability.[19] Ford was therefore given the opportunity to settle for $434,000; for Dow Corning, the figure was $366,000.

Ford and Dow Corning objected that they had not been notified earlier of their status as PRPs and should not be penalized for failing to enter an agreement of which they had no reason to be aware.[20] Both Ford and Dow Corning offered to settle for $128,000, a figure representing 4 percent of the original $3.21-million consent decree.[21] The two companies refused to join the consent decree on the government's terms, and when the government rejected their offer, they filed formal objections to the entry of the decree. The original owner-operators of the landfill also failed to join the consent decree.

Several loose ends remained as we concluded our research. Most significantly, the objections of Ford and Dow Corning had not been resolved and the consent decree had not been formally entered. Cross claims had been filed, although little effort was made to pursue them vigorously pending a decision on the decree. The operations-and-maintenance-costs issue remained unresolved. But since the court had not yet ruled on the objections filed by Dow Corning and Ford, the settling PRPs had not paid any funds to the government. In the meantime, the action against the recalcitrants was put on hold, along with potential cross claims for contributions.

Analysis

The case of Lee's Lane exemplifies Region IV's practice of applying a public works strategy to Superfund implementation. The site involved two removals. The second, a so-called removal, was really a remedial action clothed as an emergency response action. The region's traditional "clean things up" approach halted potentially fruitful negotiations and resulted in an immediate agency-conducted cleanup. Statutory limitations on removal authority were circumvented in the interests of an immediate engineering solution to an environmental problem.

Interestingly, in the subsequent cost recovery activity, the agency moved aggressively against the PRPs. This prosecutorial posture was demonstrated by the region's take-it-or-leave-it allocation of PRP shares made, as was the case with the original removal, in a context in which the PRPs were making substantial progress toward settlement on their own. It was also made evident by the agency's punitive orientation toward Ford and Dow Corning. This attempt to extract the last dollar while forcing a conclusion to the negotiations did not facilitate agreement. Rather, it contributed to a delay in the settlement and in the recovery of the costs expended by the agency on the cleanup.

While the EPA provided notice to the PRPs, its enforcement actions were limited primarily to debt collection following cleanup. The region apparently decided that it could best achieve its goals by rapid completion of an agency-directed cleanup; PRPs would be presented with a bill after the fact. There was no agency threat that failure of the PRPs to conclude negotiations would produce a section 106 order, with a fund-led cleanup and the likelihood of treble damages if prompt payment was not forthcoming. Regional officials may very well have been uninterested in forcing the PRPs to conduct the cleanup. As indicated above, there was a feeling in the regional office that agency-run cleanups not only are faster, but are also less expensive than those directed by PRPs. Apart from the desire to obtain faster remediation, however, there was no particular indication at Lee's Lane of the need to move to a fund-financed cleanup.

Once cost recovery negotiations were in progress, the government devised its own allocation formula. The parties agreed to this formula primarily because the sums involved were comparatively small, certainly not large enough to justify litigation. If the individual stakes had been substantially higher, PRP behavior might have been substantially different. Indeed, the refusal of Ford and Dow Corning to accept an arguably punitive share allocation, even in the context of a relatively low absolute contribution, indicates the potential for PRP intransigence brought about by a perception that agency demands are unreasonable.

Lee's Lane appears to be a site where a public works approach was appropriate—at least if one is prepared to bend the statutory restrictions on removals. The promise of expedience was certainly fulfilled in this case: thirteen months after conclusion of the ROD, the site had been cleaned up. But if cost recovery considerations are factored in, the picture becomes more complicated. Negotiation of operations and maintenance cost requirements meant that more was at issue than merely collecting a bill for already-expended costs; the negotiations associated with that issue helped delay final resolution of the case. Although most of the actual cleanup costs associated with the site were to be repaid by the settling parties upon finalization of the consent decree, a number of pending issues, requiring time and transaction costs to resolve, held up final entry of the decree and reimbursement of the fund.

In the final analysis, Lee's Lane suggests that a public works approach can indeed produce speedy cleanup. But if the agency is interested in obtaining reimbursement for the costs of that cleanup, the effort can be encumbered by many of the problems that beset the prosecution and accommodation strategies. Even in this context of relatively low cleanup costs, and enough PRPs to make individual shares manageable, resolution of the inevitable legal issues can be time-consuming and expensive. Enforcement costs and prejudgment interest continued to mount at Lee's Lane: what began as a cleanup costing less than $3 million approached a final bill of $5 million as of 1991, a significant amount for a project viewed essentially as a removal.

The narrow scope of the remedy at Lee's Lane suggests another potential limitation to the public works approach. At landfills that do not pose an immediate health hazard, there is always the question of whether or not the relatively inexpensive remedies applied in the present will produce a need for further remedial action. The Lee's Lane remedy was limited to removal of drums and modest efforts to contain hazardous waste. The fill was not capped to RCRA standards, nor was there any serious attempt to control for groundwater contamination. It would have been virtually impossible to undertake the next most-protective remedy, estimated to cost over $40 million, in the context of a removal. Yet agency refusal to conduct the test drillings that would produce reliable information on the nature and amount of the material in the site (a decision made, oddly enough, because of alleged fears for the public health if such drillings were conducted) meant that the site might pose substantial future problems.

Davis Farm

The Davis Farm case involved the removal of wastes buried by Southeastern Waste Treatment (SWT) on the Davis family farm in Nickelsville, Georgia. The site is in a rural area in the northeastern part of the state, known for its many carpet mills. Roy Davis and his brother, Jackson, were employees of SWT, located in nearby Dalton. They agreed to have waste materials from SWT placed on their property after the company had assured them that the Georgia environmental authorities had approved the action. The site was 200 yards from the Davis farmhouse, on a slope draining toward the Coosawattee River. The river is a source of drinking water some distance downstream.

We categorized this as a simple case because it involved few PRPs, an inexpensive remedy, and no serious technical problems. However, as in many Superfund cases, the simple was made complex by other factors. Here, the EPA ended up dealing with individuals who ultimately were prosecuted under the criminal provisions of RCRA. Indeed, the EPA was brought into the case only after a criminal investigation that followed an attempted bombing of the SWT's offices, carried out to destroy records of illegal waste-dumping. In the end, the wastes at Davis Farm were cleaned up in a fund-financed removal. Despite a variety of problems, Region IV eventually recovered all of the costs of the cleanup, and most of its enforcement costs as well.

Summary of Key Events

Southeastern Waste Treatment began operation under Georgia permits in 1974, as a nonhazardous waste treatment facility. Its original purpose was to treat latex wastes from the carpet industry. In 1977 the firm, seeking to become a regional hazardous waste treatment facility, built an incinerator, drum-handling facilities, and a tank farm. In addition to its Georgia permit, SWT received an EPA interim status permit under RCRA. In 1981, SWT was purchased by Alavesco, Inc., a hazardous waste treatment company doing business in the region.

Two months after the Alavesco takeover, SWT negotiated agreements with several large generators of hazardous waste. SWT agreed to begin incinerating hazardous waste from the Anniston Army Depot in October 1981, despite the fact that it did not yet have an incineration permit from Georgia authorities. The incinerator was too small and inefficient to process the type and volume of material SWT was receiving from the army. At one point, the firm resorted to flushing the waste into the Nickelsville municipal sewer system, while claiming to the state and the army that the material had been incinerated.[22] Shortly afterward, an inspection of the incinerator indicated that it was not functioning properly. Georgia also took action against the firm because of the unauthorized incineration of the army wastes.[23]

Although clearly in trouble, SWT continued to market its toxic-disposal service. The firm's marketing, in contrast to its waste disposal efforts, was remarkably effective: SWT received ten times more waste in the first six months of 1982 than it had taken in during the same period in 1981. The firm was granted an increase in permitted handling capacity from 10,000 drums of on-site storage and 170,000 gallons of tank storage, to 25,000 gallons of drum storage and 319,500 gallons of tank storage. The quantity of untreated wastes was rapidly increasing, however, because of difficulties with the incinerator.

Problems mounted throughout 1982. SWT was cited for having ten times more drums on-site than was permitted. With that escape hatch closed, SWT rented space from a carpet warehouse and stored approximately one thousand drums of waste there, without a permit and without notifying local emergency response agencies of the presence of the waste. The company then tried mixing hazardous wastes with latex and dumping it at the local county landfill.[24]

Finally, SWT officials persuaded an employee, Roy Davis, to let the firm dispose of material on his farm. Davis was not told that hazardous waste would be put on his property, and it was suggested that Georgia authorities agreed to the proposal. From July to September 1982, the firm transported approximately one hundred truckloads of waste to the Davis farm, where it was pushed by bulldozer into ravines and covered with a thin layer of dirt.

In the end, one of SWT's officers tried to eliminate the firm's records by firebombing its corporate offices, but the bomb—like the incinerators—malfunctioned. The U.S. Bureau of Alcohol, Tobacco, and Firearms provided information on the firm to the EPA, which launched its own investigation. The U.S. attorney initiated criminal proceedings against two of the corporate officers, Everett Harwell and Eugene Roy Baggett.[25] During the criminal inquiry, the EPA learned of the wastes deposited at Davis Farm. Georgia authorities took samples in early 1984. They found 130 chemicals at the site, 30 of which were considered hazardous substances under RCRA.

THE DAVIS FARM CLEANUP. Region IV filed an action memorandum for a $450,000 removal in October 1984, explaining in the process that it did not employ a section 106 order because of the pending criminal action.[26] However, no indication was given as to why the site could not be stabilized until a PRP-led cleanup could be pursued. The removal was done in November, with the hazardous material shipped to a RCRA facility in South Carolina. The total cleanup cost came to $745,000.

Cost recovery for the Davis Farm cleanup was complicated by the three-year statute of limitations in CERCLA, and by serious doubts about the adequacy of the evidence.[27] The EPA's legal position was somewhat weak, but Region IV indicated a willingness to litigate its position. The PRPs apparently decided that the amount in contention was not worth the cost of litigation.

SETTLEMENT. Letters were sent to forty-nine PRPs, requesting information. Based on SWT records, data obtained in response to the letters, and some judicious guesses, the Cost Recovery Unit put together a volumetric share for the Davis Farm site. Thirty-eight PRPs received demand letters. The total costs sought by the EPA in cost recovery amounted to $800,000.

The PRP situation was complicated by the inclusion of federal parties in the group,[28] plus a third-party waste broker, Aqua-Tech, Inc. Because of concerns about the special problems posed by federal PRPs, Region IV sought a separate federal PRP steering committee. The federal PRPs ultimately settled for $165,000, basing the amount on a formula produced by their steering committee. The private-sector PRP steering committee used the EPA's share list as a starting point to arrive at an allocation that added up to $618,000. Aqua-Tech eventually negotiated a separate agreement with the EPA for $40,000.

The volumetric-share information provided by the EPA made calculation of the individual apportionments relatively simple.[29] Still, it was the overall dollar figure that paved the way to agreement. The amounts involved provided PRPs with little incentive to contest the issues. The consent decrees were filed in July 1989 and entered the following October. The various consent decrees covered the entire cost of the removal and all but $106,300 in enforcement costs, left to be recovered from recalcitrants.

Analysis

Waste was removed from Davis Farm to a nearby RCRA facility (as was done in the case of Lee's Lane). Region IV knows how to do removals; faced with a choice between doing a removal or stabilizing the site, waiting for the criminal action to sort itself out, and then moving against the PRPs in a classic remedial action, the region chose to go with the familiar. Unlike in the case of Lee's Lane, however, the costs of the removal were well under the statutory limits. The existence of an imminent threat was less evident, however.

Cost recovery followed the removal, although the government's tardiness and lack of early preparation of share information suggest that the decision to seek cost recovery from the waste generators was something of an afterthought. (Region IV's cost recovery unit had only recently been established.) The SWT officials were under criminal prosecution, the owners of the farm were given immunity, and, with the exception of the federal PRPs, most of the responsible parties were held accountable only for limited amounts of waste.

The success of the actual cleanup was based on the region's ability to get the site within the statutory limitations on removals, on the availability of approved RCRA facilities to accept removal waste, and on the suitability of the site to a removal-style cleanup. The success of the subsequent cost recovery, as at Lee's Lane, was related to the relatively low cost of the cleanup for the various PRPs associated with the site. We heard from PRP lawyers in this case many of the same refrains we heard from lawyers for the Lee's Lane PRPs: while there were real questions about liability, about the proportional shares the government worked out, and about legal standards being applied in the case, the low amounts at stake simply did not justify the transaction costs that would accompany formal resolution of the issues. Again, this aspect of the case suggests that while the public works approach can produce an expeditious removal-style cleanup, the success of subsequent efforts to recover costs from PRPs is neither assured nor likely to produce quick results, particularly at larger, more complex sites.

The Davis Farm case suggests that if the costs to individual PRPs are low enough, almost any enforcement regime can be successful, both in getting sites cleaned up and in minimizing costs to the government. Given the government's legal and evidentiary problems, this case presented PRPs with as potentially strong a case against CERCLA liability as any encountered in our research. In addition, the fact that many generators sent their waste to SWT in good faith, precisely because it was a RCRA-approved facility, suggests that a heavy-handed application of joint and several liability may not have been entirely appropriate. Yet these arguments against saddling PRPs with cleanup costs that were arguably the result of inadequate regulatory performance by state and federal officials were not seriously advanced by the PRPs in the Davis Farm case. It would appear that the PRPs made a cost-benefit calculation and decided to pay the cleanup costs rather than absorb the predictably higher transaction costs that would accompany resolution of the legal issues in court. That fact, plus close cooperation and coordination among the various steering committees, facilitated prompt settlement.

The Davis Farm case, like Lee's Lane, demonstrates the utility of government help to PRPs in allocating cleanup costs among themselves. SARA authorizes the EPA to compile nonbinding allocations of responsibility (NBARs), if asked by the PRPs. However, the device is almost never used.[30] Region IV's use of EPA-compiled volumetric shares is a kind of informal NBAR; such information serves many of the same purposes served by NBARs, but the EPA can produce it without a PRP request. Volumetric-share data, when the agency is willing to prepare and disclose it, appears to provide strong support for the formation of steering committees and ultimate settlement.[31]

Assessing the Public Works Strategy

The public works strategy assumes that cleanup can best proceed through immediate government action at a site, followed by efforts to recover the costs from the PRPs. Of the three implementation strategies available to the EPA, the public works approach includes the fewest conceptual ambiguities. No assumptions are made about PRP behavior. The agency simply is empowered to do what it knows how to do best: remove toxics from sites, take them to an approved disposal site, and send the bill to the PRPs.

Region IV officials behaved in accordance with the dictates of the public works model in both the Lee's Lane and Davis Farm cases. The engineers remained in control during the cleanup phase. Lawyers subsequently stepped in as bill collectors. The desire to take aggressive cleanup action is illustrated by the region's decision to terminate potentially productive PRP negotiations in the Lee's Lane case and proceed with a removal action, and by the immediate recourse to a removal in the case of Davis Farm, despite some doubt about the presence of an imminent hazard. The bill-collector role cast for lawyers in the public works strategy was also assumed in both cases, although an excess of prosecutorial enthusiasm almost derailed the settlement in Lee's Lane, and the legal issues raised in Davis Farm almost certainly would have produced more resistance if the overall costs of the cleanup were much higher.

How well did the public works strategy fulfill the goals of the Superfund program? A speedy cleanup was achieved in both Lee's Lane and Davis Farm. Public expense and transaction costs were kept low, although the situation may have been otherwise if higher overall costs had led the PRPs to contest the cost recovery actions of the EPA. The issue of the appropriateness of the remedy is more difficult to assess. In both cases the agency imposed low-cost, low-tech solutions. No impermeable cap was put over either site, despite the potential for groundwater problems at both, nor were there any efforts to clean up contaminated soil. The agency's admission that it would not make test drillings at Lee's Lane because of its fear that the very act of testing might produce dangerous results does not augur success in achieving the permanent cleanup mandated by the statutes.

The remedies applied in these two public works cases also illustrate what is perhaps the major limitation of the public works study. Hazardous waste cleanups are expensive—the average remedial action now costs more than $30 million. The removial tool, restrained as it is by a $2-million cap on expenses for any one action, is inadequate for sites of such magnitude. While Region IV had become adept at stretching this constraint by a variety of means, such activity has practical limits.

The public works approach to Superfund cleanups, in the context of the expense limitations, inexorably presses government officials to choose remedies that can be applied relatively cheaply. A concern over costs is entirely appropriate in Superfund, but as the extent of environmental damage and public health risks increase, a relatively low ceiling on cleanup expenses may tilt the balance too far in the direction of "quick and dirty" engineering solutions that do not address potentially serious environmental problems at a site. We, and some of our interviewees, suspect that the remedy applied at the Lee's Lane site may fit into this category.

Assessing the success of different implementation strategies is best done comparatively. In the following chapter, we move from the particular to the general, from the individual strategy to a comparative analysis of all three. We also assess the degree to which the various approaches to Superfund fulfill the different and sometimes conflicting goals of the program.

Part Three
Cleaning Up the Mess

Six

What Works?

THE SUPERFUND program was designed to effect the decontamination and remediation of the nation's inactive toxic waste dumps, primarily through the efforts and resources of the private businesses whose prior actions at a particular site made them legally responsible for its cleanup. Despite this superficial clarity of purpose, the Comprehensive Environmental Response, Compensation, and Liability Act of 1980 (CERCLA) and the Superfund Amendments and Reauthorization Act of 1986 (SARA) established what is probably the most complex statutory scheme in the history of environmental legislation. It is certainly unique in the number of unfamiliar and largely unguided implementation choices presented to the agency charged with its administration. Further, the enormity of the statutory goal—in terms of the costs involved in cleaning an individual site and the number of sites in need of attention—was unforeseen by the architects of the legislation.[1] Under such circumstances, the initial confusion and subsequent disparate responses of the Environmental Protection Agency should come as no surprise. Neither should the resulting tensions within the EPA, and between that agency and Congress, the Department of Justice, state environmental agencies, and the private businesses charged retroactively with the responsibility for the cleanups.

In Superfund's first decade, few sites were completely cleaned up and subsequently "delisted," although a growing number entered the phase of remedial action.[2] At these sites most of the action was being taken by bulldozers, rather than lawyers and engineers, because the basic decisions had been made concerning design of the remedy and the allocation of responsibility for its execution and financing.

As the preceding chapters have shown, the period since the 1980 passage of CERCLA has witnessed the application of a variety of Superfund implementation strategies. We have taken advantage of that variation to examine the implications of these differing approaches. In the most basic terms, we are interested in providing some tentative answers to that central question, what works? The comparative dimension of this project also allows us to address the frequently irreverent rejoinder to that question, compared with what?

To be useful to policymakers, answers to these questions must take into consideration the context in which the EPA operates. Superfund implementation strategies are a product of the agency's efforts to cope with a series of imperatives:

—Legal imperatives are posed by the multiple and competing goals established in CERCLA and SARA as interpreted by the courts, by complex procedural safeguards and substantive standards set by these statutes and others of more general applicability, and by the set of specific statutory tools provided by Congress to the EPA for cleaning up sites and dealing with potentially responsible parties (PRPs).

—Resource imperatives are driven by the program's finite monetary resources, the use of which is limited to a set of statutorily defined purposes. Resource imperatives are also shaped by the personnel and organizational structures of the agency, and by the constellation of consultants and contractors available to aid in the agency's activities.

—Bureaucratic and political imperatives are the result of demands placed on the agency by Congress, the White House, state and local government agencies, and the general public. These imperatives are also the result of the EPA's need for the cooperation of outside organizations such as the Department of Justice and the courts if it is to make use of its enforcement tools.[3]

In addition to these broad and relatively fixed elements of the policy environment, Superfund implementation is shaped by a set of site-specific conditions that are also subject to little alteration: the physical features and location of the site, the nature and extent of contamination, the state of records and evidence, the amount and nature of public attention, existing relations with the relevant state agencies, and the number and nature of PRPs associated with the site. We begin with the assumption that toxic waste site cleanups entail some irreducible amount of managerial and administrative work, a demanding inter- and intra-governmental clearance process, substantial scientific and technical efforts, much uncertainty, and the need for someone to expend an undefined—though frequently very large—amount of money.

As is argued in chapter 2, the policy environment compels the EPA to rely primarily on Superfund's liability scheme to achieve agency objectives in the program: in each implementation strategy we have posited, the responsible parties are expected to take responsibility for a substantial share of cleanup costs. The strategies differ not so much in their grounding in PRP liability for cleanup, but rather in the procedures by which the government applies the liability scheme, and the degree of flexibility demonstrated in that application. The product of policy analysis is advice.[4] Given a choice of strategies in a context of constrained agency action, what advice can we offer to policymakers? What works? Or, what might be expected to work best in particular situations?

Examining the Approaches

In chapter 2, we set out the four goals we believe should inform an evaluation of the Superfund program: *appropriateness of the remedy, minimization of taxpayer expense, speedy remediation, and minimization of transaction costs.*

We suggested that these goals are ill-defined and are at least potentially in conflict. Not surprisingly, the EPA has not grasped the political nettle and set out a comprehensive framework in which the trade-offs inherent in these objectives can be made explicit. As a result, balances among competing programmatic goals must be struck anew at each site, and personnel at the regional level have tremendous discretion in how to define their objectives.

How do the prosecution, accommodation, and public works strategies compare in their achievement of the four goals listed above? Does each strategy imply a distinct, if implicit, set of trade-offs among Superfund's competing goals? What operational lessons can be derived from the case studies? In this chapter we address these difficult questions. We first discuss, in turn, each of the four objectives of Superfund, and assess the comparative performance of the implementation strategies in achieving it. We conclude with some broader lessons for implementation of the Superfund program.

Appropriateness of the Remedy

The adequacy of the remedies selected at Superfund sites has been the subject of much discussion and criticism.[5] A complete assessment of success at choosing remedies would require a site-specific determination of that central question: How clean is clean? Or, more to the point at an individual site: How clean is clean enough? Highly general cleanup standards are described in the law: a cleanup must be "protective of human health and environment"; it should embody a preference for "permanent" solutions, and be "cost-effective." These directives are vague and—depending on how cost-effectiveness is defined—in conflict. More important, they beg the most serious questions facing policymakers at Superfund sites: *How* protective? *How* permanent? *How* cost-effective?

Complicating the situation is uncertainty over scientific and technological issues: the acceptable levels of risk to health and environment associated with each remedial alternative and the availability of reliable measures of that risk, the likely effectiveness and permanence of the various engineering options, and the probable cost of these options. As a result of this compounded uncertainty, decisions among the leading alternatives that emerge from the remedial investigation and feasibility study (RI/FS), and choices concerning which cleanup standards (or ARARs) will be met, are driven more by political than by technical criteria.

Determining if one remedy is superior to another is not possible without a rationale for judging either the quality of outcomes or the quality of the process itself. No one has developed a widely accepted criterion for judging the quality of outcomes at Superfund sites. Differences among the leading remedial alternatives almost always concern degrees of risk reduction; choices rarely are clear-cut. Previous studies have assessed the remedies selected at Superfund sites in terms of some fairly elemental criteria such as cost or generic type of remedy selected.[6] Although we are reasonably familiar with the remedies selected at the sites examined in the case studies, we are not competent to assess their adequacy in a technical sense and are uncomfortable in applying the rough-and-ready indicators of previous studies.[7]

As indicated at its outset, this book is concerned more with the process of decisionmaking at Superfund sites than with the substance of remediation decisions. Because of this orientation, as well as the difficulties involved in assessing the technical adequacy of remedies across a number of sites, we examine in this chapter the sufficiency of the remedy from the perspective of process. We posit—plausibly, but admittedly without much empirical support—that a process that encourages participants to consider and weigh all the relevant factors, that promotes the generation of reliable technical data, and that ensures meaningful participation by all the relevant stakeholders at a Superfund site will produce better remedies than alternatives that do not encourage these procedures.

Clean Sites, a nonprofit organization dedicated to improved decisionmaking at Superfund sites, recently issued a comprehensive report on Superfund remedy selection that offers recommendations aimed at making the process of choosing among remedial alternatives "more explicit, more reliant on the interaction among the various interests, more transparently logical, and more consistent."[8] Two procedural issues examined in that report are of special relevance to our own research: stakeholder (particularly PRP) participation in remedy selection, and the temporal ordering of decisions relating to the type of remediation specified and the substantive standards of cleanliness to which it must conform.

STAKEHOLDER PARTICIPATION. In our research, we found that each of the three implementation strategies had a distinctive effect on the degree of stakeholder participation. The remedy selection process in the prosecution strategy was driven by the minimum-disclosure requirements of the relevant statutes, and by the frequently instinctive reluctance of legal professionals to reveal anything voluntarily in an adversarial setting. Stakeholder participation was therefore limited. In Laskin Poplar, the complex Region V case in which a prosecutorial strategy was applied (see chapter 3), remedy selection was closed, even secretive. Even in the Cliffs Dow case (see also chapter 3), in which the PRPs conducted the RI/FS themselves, the process was characterized by acrimony and an adversarial, legalistic posture on the part of the government. There was a rough equal-ity of participation opportunities, however. PRPs, the public, and citizens' groups all were permitted to speak only on statutorily defined occasions, and all had access to the same set of limited data.

These observations emphasize that, when a prosecution approach is taken, the government's concerns are paramount and those of others are important only insofar as their cooperation is required for the government to achieve its ends. The degree of influence that others have over the process of remedy selection in the prosecution strategy thus is defined by whether or not the government needs their input, rather than by what they have to say.

The remedy selection process was much more open in the cases we examined in Region III, in which an accommodation strategy was used. (See chapter 4, the cases of Tybout's Corner and Harvey and Knott Drum.) Technical information was readily shared by the agency as it became available, and PRPs had direct input into the remedy ultimately set out in the record of decision (ROD). This approach to the RI/FS produced remedies with which both the EPA and the PRPs were comfortable.[9]

The cases in Region III suggest that sharing technical information and allowing PRPs to participate in remedy selection, while requiring more agency effort in the short term, may result in more effective decisionmaking in the long term. The routine and systematic sharing of information between the EPA and the PRPs in Region III cases, and the collegial nature of decisionmaking in the feasibility studies and subsequent RODs, served to increase the EPA's information base and its range of potential remedial options.[10] If the literature on the determinants of compliance with court orders is applicable in this context, this increased participation by PRPs may have augmented their satisfaction with the process and led to fuller compliance with the terms of the consent decrees.[11]

We note, however, that the increased participation of PRPs in the accommodation strategy may come partly at the expense of participation opportunities open to other publics that are more sporadically involved. The local community and other citizens' groups were uninvolved in the informal cleanup negotiations in Region III; they had only the formal participation opportunities required by law. Public interest typically peaks at the time of the hearings on remedies, prior to the issuance of the ROD. In Region III, the effectiveness of such arenas was compromised by the fact that many important decisions had been made earlier, in less formal negotiations between government officials and PRPs. Furthermore, the preference at these formal decision points was for broad decisions, with details to be worked out informally as remediation progresses.[12]

The public works approach typically involves even fewer opportunities for stakeholder participation in remedy selection than does the prosecution approach. The emphasis is on speed, and participatory decisionmaking processes are seldom expeditious. An RI/FS is not required in removal actions such as Davis Farm, nor are formal mechanisms for public participation prior to issuance of the ROD.

THE SEQUENCE OF DECISIONS. Clean Sites' report on remedy selection asserts that "EPA often explores in-depth all the alternative cleanup methods it plans to consider before defining its cleanup objectives and the level of protection it is seeking."[13] We observed this sequence of decisions in each of our case studies. This approach led to confusion concerning the criteria to be used for remedial choice, disagreements among the various parties over how to apply the various possible cleanup standards, and acrimonious and time-consuming disputes over key terms such as "permanence," "cost-effective," or even "protection of human health and the environment." Arranging decisions in this order pushes some of the most difficult and controversial issues to the end of the process, to be managed primarily by the PRPs.[14] Both Region V cases, for example, involved protracted discussions of the specific applicable or relevant and appropriate requirements, the ARARs that would define remedial success. These discussions took place well after the RODs were issued. Indeed, the final discussions over state sign-offs occurred after the consent decrees were signed.[15] This process allowed a certain amount of flexibility and informality in regard to ARARs requirements, but the overall goals of the cleanup necessarily were circumscribed by what the chosen remedy could be expected to accomplish.

The reasons behind the prevalence of the current approach are not difficult to ascertain: final cleanup levels are a subject of controversy and the statutes provide little concrete assistance to the EPA in specifying them; the various parties to Superfund negotiations typically have much at stake in where the levels are set; and it is often especially difficult to set cleanup objectives at an early stage in the process, when knowledge of the nature of contamination on the site, and the likely impact of remedial measures, is least clear.

Clean Sites has recommended that the agency reverse the order in remedy selection, and establish site-specific cleanup objectives and contaminant levels *before* remedial alternatives are developed. These objectives and levels would be set through "an explicit and interactive process" in which all relevant stakeholders at the site would participate: the EPA, state environmental officials, PRPs, and representatives of the local community. An important aspect of this proposed process is disclosure by the EPA of "all verified site information as soon as it is available."[16]

Forcing an early resolution of the contentious issue of cleanup objetives and containment levels may delay the actual cleanup and raise the overall level of conflict.[17] Yet clashes over these standards probably are inevitable, and the only choice may be to select the point in the process when their resolution will be least costly, and most conducive to the choice of an appropriate remedy. If our cases are at all typical, lawyers and legalistic arguments most often come to the fore at the end of the decisionmaking process, when consent decrees are being negotiated and, frequently, threats are being exchanged. Cleanup levels and questions of ultimate land use are scientific, economic, and—most important—political in nature, rather than legal. The course of prudence may be to resolve these questions at an earlier stage, when lawyers and legalistic considerations may have less impact on the process.

A NOTE ON CLEANUP COSTS. The likely overall cost of a remedy influences all procedural aspects of remedy selection, as suggested most clearly by the comparative ease of decisionmaking in the simple, less costly cases examined under all three strategies. We do not equate cost with quality of remedy, although some critics of Superfund have used cost as a surrogate of quality.[18] However, we can offer some preliminary observations on the relationship between implementation strategy and the cost of the remedy selected.

With the prosecution strategy, there is a tendency for government to state a preference for the more environmentally protective—and usually more costly—remedies, at least at the outset of the remedy selection process. However, one should keep in mind that this behavior is sometimes a means of establishing bargaining positions for negotiations at the end of the process. In both of our prosecution-strategy case studies, the remedies ultimately chosen represent compromises between the more expensive solutions favored by government and the less expensive ones advocated by the PRPs.

When the accommodation strategy works properly, the remedy selection process operates throughout to encourage compromise. Two factors may prevent the more expensive solutions from giving way to the necessity of obtaining agreement from PRPs: the prospect of mixed funding (if the government wants an especially expensive solution, perhaps it should be willing to expend some of its own resources to pay for it), and the possibility of suing recalcitrants and other nonsettlers for the difference between the cost of the government's preferred remedy and the last best offer of the cooperating PRPs.

Our analysis of the public works strategy, however, shows quite a different result.[19] Here we found a disquieting tendency to consider seriously only remedies that met the $2-million statutory limit for emergency removals. In such cases, quick action may have been purchased by a choice of remedy that was less costly—and less protective—than one that would have been chosen under the other strategies. In that sense, then, the public works strategy may blunt an important characteristic of the current liability system: a tendency for government to consider the requirements of protectiveness over considerations of cost.

Minimizing Taxpayer Costs

The Superfund scheme, despite a nomenclature that seems to emphasize fund-financed actions, was clearly designed to compel the legally responsible private parties to pay for cleanups. The various approaches to implementation therefore must be assessed in light of this central objective. However, it is important to disaggregate the costs that may accrue to taxpayers in a Superfund case. While the categories of such costs overlap substantially, they raise distinct issues for different participants in the process.

There are, first, the actual *cleanup expenses:* the scientific testing and engineering design, the construction costs, and the expenses involved in an ongoing remedial program. In the case of removals or remedies that take contaminated material off site, there are also transportation and disposal expenses.

Administrative costs constitute another class of government expenses: handling relations with the local community and with other state and federal agencies, and supervising contractors and consultants through all stages of remediation.

A third category of cost is related to *future risks*. There is the ever-present possibility of finding, in the midst of remedial action, a far more serious problem than was originally contemplated, or of ascertaining that a carefully crafted remedy is ineffective. These risks imply potential costs that may be borne by the PRPs, or the Superfund, or the state, or some combination of the three. We have observed that one of the most contentious issues in Superfund negotiations is allocation of these risks.[20]

PRPs are often most concerned about future risks, and are willing to pay a substantial premium over a "fair share" of predicted cleanup costs in exchange for certainty about the size of their final burden. The specter of limitless liability for the unknown and unknowable problems that may develop at a Superfund site, decades in the future, is profoundly disquieting to corporate managers. Recognition of this fear underlies the *de minimis* provisions of the statutory scheme. The desire of corporate PRPs to obtain some degree of closure on as many aspects of future risk as possible was present in nearly all the cases we examined. It underlies the pressure to obtain covenants not to sue from the government, and the reluctance of PRPs to accept consent decrees with broad reopener provisions. It is also a key factor in the PRPs' desire to make the consent decree and remedial work plans as specific as possible, so that their obligations might be clear and their costs controlled.

Superfund's liability scheme gives the government a strong position relative to the PRPs on the issue of cleanup costs. So long as reasonably persuasive evidence links one or more private parties to a Superfund site, and at least one of those parties has sufficient financial resources to foot the cleanup bill, the tools exist to make that party assume responsibility for all the costs associated with a cleanup. It would appear that the government need only be tough enough, wait long enough, and impose on itself and the other parties the legal and other transaction costs necessary to obtain the desired result.

The fact that the government did not achieve 100-percent solutions in any of the cases we examined indicates that the government's position in Superfund cases is somewhat more complicated than the above scenario suggests. If the threats, orders, and panoply of other coercive devices fail to achieve "voluntary" PRP assumption of all the administrative, cleanup, and risk costs, then the government must be prepared to litigate and force compliance. And, while the federal government holds most of the legal cards in this judicial showdown, it still must convince a federal judge that it is acting both within the law—the easy part—and reasonably. The requirement of at least the appearance of government reasonableness emerges not from the statutory language of Superfund (which is arguably unreasonable), but from what might be termed judicial imperatives.

Federal judges have crowded dockets. From the cases we have studied, and from reports of others, it is clear that the last thing most federal judges want is to preside over a big Superfund case. The large, complex CERCLA cases (the only ones that may be worth litigating) typically involve scores of PRPs, complex and conflicting scientific and engineering data, and evidence that is problematic at best. The consequence is that the judges inevitably press both sides to settle. And settlement normally implies *mutual* concessions. This dynamic, at least in the cases we examined, impels both sides to "split the difference," to "give a little to get a little." The unremitting judicial insistence that both sides negotiate, together with the political and bureaucratic pressures brought on by delays and mounting legal costs, leads almost unavoidably to agreement by the government to a final resolution of the case that is less than the complete victory that the liability scheme would seem to promise to those pursuing a pure prosecution strategy.

The accommodation strategy emphasizes the mutuality of interest between the government and PRPs. It treats the PRP more as a responsible corporate citizen than as an "amoral calculator" bent on maximizing profit at the expense of the public interest.[21] In this context, notions of fairness and equity play a bigger role than in the prosecution model. Many of the provisions in SARA that support the accommodation approach to Superfund implementation suggest that, in appropriate circumstances, *some* of the costs of cleanup should be borne by the government rather than by PRPs.

These cleanup costs may be administrative, as in government forgiveness of past administrative costs, or preparation of an NBAR (nonbinding allocation of responsibility) or analogous volumetric-share document. These costs may include some of the actual remediation expenses, covered by means of either implicit or explicit mixed funding, as in the case of Harvey and Knott, and the case of Tybouts Corner. The government might arrange to cover costs by assuming some of the future risks inherent in the cleanup endeavor. The rationale for these concessions, in light of the tools at the government's disposal to compel PRPs to assume *all* costs, can include reducing delay in getting a site cleaned up, minimizing transaction costs, or even increasing fairness to one or more of the PRPs. Whatever the rationale, the accommodation strategy does not promise to keep the government's share of cleanup costs to the absolute minimum.

While the prosecution strategy may be an effective means of extracting somewhat more in the way of administrative, cleanup, and risk costs from reluctant PRPs than alternative strategies, the difference may not be as great as expected. If negotiated resolution of a Superfund case is inevitable, then the question becomes not so much *whether*, but *when* to negotiate and compromise, and over what. In the final analysis, use of an accommodation strategy may not cost the government significantly more than the prosecution alternative.

The public works approach is more difficult to categorize in terms of ultimate cost. The major element of the public works approach to Superfund cleanups is up-front expenditure of Superfund monies to clean up sites. Reimbursement for government outlays subsequently are sought from PRPs in cost recovery actions. The concern is to get a site cleaned up as quickly as possible, and let the lawyers figure out at some later time who will pay the bills. As practiced in Region IV, the public works approach also has involved agency effort to obtain accurate information on the PRPs at a site, and their respective contributions of waste. This effort incurs administrative costs that are eschewed in a prosecution approach, in which the expenses involved in identifying additional PRPs and calculating volumetric shares are left to the PRPs.

There are also risk costs involved in the public works strategy. Successful cost recovery depends on a number of factors, such as the EPA's ability to demonstrate to the satisfaction of a court that an "imminent and substantial danger to human health and the environment" existed at the site, that the remedy chosen was appropriate, and that the parties being sued are in fact legally responsible for the cleanup. There are also risks from potential PRP challenges to costs incurred by EPA contractors. These obstacles are not insurmountable, but they inevitably involve more uncertainty about the government's final share of costs than is present when, as is the case with other approaches, agreements over most costs are made before they are actually incurred.[22] The relatively low cost of removal remedies dampens the incentive to litigate in public works cases, however, as our cases illustrate.

If cost recovery is complete, then the public works approach may achieve the same objective as the prosecution approach, with faster remediation and possibly lower transaction costs. But, as we indicated in chapter 5, the public works approach is feasible only at relatively small sites, where removals and removials are possible. At large, complex, potentially expensive sites, the bill-paying approach is less likely to be used, and the real choice lies between accommodation that begins with the assumption that concessions need to be made, and a prosecution approach in which the government is forced into concessions only at the end of a lengthy, contentious process.

Speed of Remediation

A concern with speedy cleanup underlies much of the Superfund program. It is most explicit in the removal provisions of CERCLA, in which immediate action is authorized in emergency situations. But if there is a common theme in the chorus of Superfund criticism—from right and left, from environmentalists, members of Congress, and from representatives of corporate America—it has been that cleanups have been too slow. Substantial grounds exist for this criticism. By mid-1990, for example, after ten years of program operation, only sixty-three of the more than twelve hundred National Priorities List sites had been cleaned up, and only twenty-nine delisted after long-term effectiveness of cleanup was verified.[23]

The glacial pace of cleanup activity during the first years of the Superfund program led Congress to specify in the program's first reauthorization (SARA) targets requiring the EPA to begin a specified number of RI/FSs and remedial actions by dates set in the statute.[24] A concern with speed undoubtedly underlay SARA's change to administrative decisionmaking in remedy selection, as well as the substantial constraints placed on the timing and grounds for legal challenges to the remedy chosen by the EPA at a particular site.[25]

The cases examined in this book suggest that relatively prompt cleanups can take place at simpler sites, regardless of the implementation strategy chosen. But no matter how the government chooses to proceed at complex sites, remediation decisionmaking concerning technical issues is unavoidably lengthy. The RI/FS process at major sites, even when optimally efficient, can last one to three years.[26] Remedial design can take almost as long. The actual remediation can consume a decade or more, especially if the remedy selected involves "pump and treat" or similar technologies. These time requirements exist independent of implementation strategy, because of the technical and engineering questions that must be answered before remediation can begin, and because the actual physical cleanup is frequently a time-consuming process. The issue, at least with complex sites, may not be which of the various approaches to Superfund implementation promises speedy cleanup, but rather which is least likely to add delay.

That much being said, Laskin Poplar and Tybouts Corner suggest that the time needed to achieve actual cleanup may be lengthened substantially when a prosecution approach is taken to Superfund. These two cases each consumed more than a decade in legal maneuvering before remedial action began, despite the fact that neither actually went to trial.

The accommodation approach, emphasizing the common goals PRPs share with the government, and assuming a modicum of good intentions on PRPs' part, promises a reduction of legal wrangling and a resulting acceleration of the cleanup process. The remedy selection process in the two cases we examined in Region III lends credence to this proposition. It took little more than a year from the time that the PRPs became active in the technical decisionmaking process until the EPA and the major PRPs reached agreement in principle on the broad issues of remedial design.[27] Technical and legal issues in Harvey and Knott Drum were similarly resolved in slightly more than a year from the time that General Motors became actively engaged in the RI/FS process.

The two Region III cases cast doubt on the proposition, heard in several of our interviews with government officials, that application of the statutory provisions for PRP participation in initial technical decisions leads inevitably to delays. Rather, our research suggests that it is the prosecutorial orientation that tends to provoke resistance, wrangling and, therefore, delays. It may be that the government and its contractors can complete the RI/FS and ROD more quickly and efficiently without any PRP participation whatever. But if responsible PRPs are given an opportunity to participate in a meaningful dialogue with government officials concerning the remedy selection decision, subsequent stages of the process are less likely to bog down in legal and technical arguments over the acceptability of the chosen remedy. This is the underlying premise of the Clean Sites recommendations concerning remedy selection, which explicitly advocate movement toward a more interactive process.[28]

Not surprisingly, the public works approach proved to be most conducive to speedy—though not necessarily thorough—cleanup in the cases we examined. The cry of "bring in the bulldozers" that underlies this implementation strategy derives from the frustration one would expect of an engineer faced with the seemingly interminable and unproductive delays brought about by legal and technical squabbles between lawyers and bureaucrats. The problem is that not all Superfund sites can be squeezed into the procrustean bed bounded on one side by $2-million aggregate cleanup costs and on the other by a finding of imminent and substantial endangerment.

When a public works approach is statutorily or bureaucratically unfeasible, the alternatives are prosecution or accommodation. Harvey and Knott and the early, technical phases of the Tybouts Corner case suggest that the attempt to involve PRPs from the beginning, through a cooperative approach in which the government attempts to meet the private parties part way in terms of remedy selection and the apportionment of costs, can advance the process more quickly than more aggressive, legalistic procedures.

Minimizing Transaction Costs

Transaction costs as defined here—the costs that accompany the process of determining financial responsibility for a cleanup—are related to speed of remediation: typically, transaction costs mount as delays increase. The one situation in which a speedy remediation does not necessarily mean lower transaction costs is when the public works approach is followed by cost recovery actions. As both Region IV cases indicate, cost recovery *after* completion of the actual cleanup sometimes can result in substantial legal costs as cleanup expenses are allocated among the PRPs through either negotiation or litigation.

Most transaction costs are related to legal expenses, because it is lawyers who negotiate or litigate the issues of PRP responsibility at individual sites. Because the basic element of CERCLA's statutory scheme is assignment of legal responsibility for cleanups to private parties, through the courts if necessary, some transaction costs are inevitable, at least when viable PRPs exist at a site. Transaction costs in Superfund, however, are extraordinary by almost any standard. They have been estimated to range between 20 and 40 percent of the aggregate costs incurred by private parties at a typical site. With the average cleanup cost per site approaching $30 million, the aggregate societal resources being committed to decisionmaking concerning Superfund sites becomes a matter of serious concern.[29]

We emphasize that our focus here is on aggregate transaction costs, not just those accrued by the government. We also admit that we have no precise measures of transaction costs in the cases we examined, given that these data frequently are difficult for parties in Superfund cases to assemble, and tend to be regarded by them as confidential in any event. Even absent an ability to make precise assessments of these costs, the stark differences in procedures used to resolve the cases discussed in this book make general conclusions about transaction costs almost self-evident.[30]

Prosecution involves the maximum possible shifting of the costs of cleanup, administration, and future risk from the government to the PRPs. Shifting of transaction costs is also a central element of the prosecution strategy. This approach almost certainly reduces the initial transaction costs of the government. PRP searches can be costly and time-consuming, despite the broad investigative powers conferred on the EPA by CERCLA. Whether this form of cost-shifting actually reduces the government's transaction expenses in the long run, given the litigiousness and acrimony such procedures tend to generate, is less clear. It is almost certain, however, that this strategy increases the *aggregate* transaction costs to all parties to the process.

It is difficult to imagine a less efficient, more expensive method of determining which corporations' waste was disposed at a site than the process used in the third-party discovery procedures in the Tybouts Corner case. Much of this research almost certainly could have been carried out more effectively, at far less cost, through a government investigation conducted by engineers and administrators rather than lawyers. (The increasingly common practice in Region IV of developing "waste-in" lists for use by PRPs is an example of just such a process.) The inherently adversarial and legalistic method of factfinding employed in the cases of Tybouts Corner and Laskin Poplar also generated additional transaction costs in the form of disputes over what information was and was not relevant to the case—what was "discoverable," in a legal sense.

Finally, the prosecution strategy, when applied to difficult, multiparty cases, also created a perverse incentive that encouraged PRPs to engage in third-party litigation. Once it became clear that the government had targeted them, and that they would most likely bear considerable remediation costs, PRPs had a clear incentive to spread that burden—as widely as possible. This option took the form of suing others to bring them into the set of parties among whom costs would be spread.

Transaction costs are thus problematic in a prosecution strategy because individual incentives push them up, to the detriment of all participants. EPA officials evince little concern over the transaction costs of the PRPs. These costs are seen as self-inflicted wounds: instead of being inappropriately contentious and recalcitrant, PRPs should simply yield to the government's superior legal position. To the extent that government administrators consider their own transaction costs (and it is not clear that they do with any frequency), they view them as recoverable from the PRPs. From the PRPs' perspective, transaction costs are a necessary part of their only possible response to the legal initiatives undertaken by other PRPs, or the perceived unreasonableness of government demands.[31]

The accommodation approach to Superfund emphasizes the common interests of government and the PRPs, in particular their shared interest in reducing delays and transaction costs. Accommodation finds statutory support in a number of SARA provisions designed to ease the way to settlement. These devices have two characteristics in common: they embody departures from pure joint and several liability toward some notion of distributive justice and fairness, and they involve governmental assumption of some elements of the cleanup and transaction costs at a site. Utilization of settlement devices requires agency effort and, therefore, the assumption of some costs. In the case of mixed funding, *de minimis* buyouts, and releases from future liability, these devices also involve government payment of some portion of the costs of cleanup, administration, and risk associated with remediation of a site.

We did not examine a large, complex site at which the accommodation approach was used throughout the decisionmaking process.

We suspect that the number of such sites may be quite small because the larger sites tend to be handled (or at least closely supervised) by the Department of Justice, whose approach to Superfund cases is prosecutorial. The strategy taken toward Harvey and Knott Drum, the less complex site we examined in Region III (see chapter 4), embodied several elements of the accommodation approach.[32] Negotiations were characterized by all parties as "very smooth," even "easy." Although we have no concrete figures on transaction costs to the various parties, it seems clear that the aggregate transaction costs at Harvey and Knott Drum were far less than in either Davis Farm or Cliffs Dow, the "easy" cases from the other regions.

The public works approach as practiced in Region IV puts cleanup first, with activities to compel PRP payment coming at the end of the process, in the form of cost recovery actions. Transaction costs therefore are not eliminated, merely postponed. Deferral of transaction costs does not necessarily result in their elimination. Indeed, because PRPs are presented, after the fact, with a bill for cleanup expenses, they have both an opportunity and an incentive to contest various elements of that bill. It is only in cost recovery actions that PRPs can challenge various aspects of the remedy or the specific costs charged to them. There is no inherent reason why legal conflict, and the resulting transaction costs, should be less at the end than at the beginning of the process.

In Region IV, however, the "agency as bill collector" appears to have been successful at obtaining PRP compliance with infrequent resort to the courts. This success appears attributable to the establishment of procedures and organizational changes that are not without cost to the government: the preparation of waste-in lists, for example, requires substantial investigation; accounting practices within the agency must be defensible if the government is to be reimbursed for administrative costs; choice of cleanup procedure, and (if relevant) the requisite immediacy and seriousness of risks must be carefully documented if these agency decisions are to stand up to subsequent scrutiny. This approach is also limited primarily to those sites where removal actions are feasible.

Thus, at relatively small, noncomplex sites, a public works approach, followed by cost recovery, may minimize transaction costs, although an accommodation approach may do so to equal effect. In fact, the "bill collector" mode adopted by Region IV's cost recovery unit involves the EPA in a number of the same tasks involved in the accommodation approach: the region conducts thorough PRP searches and helps PRPs assign proportional shares of the costs through preparation of a waste-in list. The difference is the role of PRPs at the beginning of the process, when those agency staff taking an accommodation approach involve PRPs in technical decisions and make substantial efforts to gain their acquiescence in a consent decree before much on-site work is done; the public works approach, when followed by cost recovery, simply postpones this activity until later in the process. At a large site where removals are not feasible, the choice is typically between accommodation and prosecution. In these circumstances, successful application of an accommodation approach implies lower transaction costs.

Lessons

Superfund implementation, as indicated at the outset of this chapter, must be understood in the context of the imperatives that constrain the EPA's actions. These include the legal mandates established by Congress and the courts, resource constraints, and the bureaucratic and political imperatives flowing from internal and external agency relationships. It is within this framework of limits that the agency must sort out the conflicting goals established in the uniquely complex Superfund program. As has been shown, this milieu of contradictory goals, diverse tools, and constraining influences has produced divergent implementation strategies, over time and across geographical regions. Our case studies were chosen to enable an assessment of the impact of these approaches, and the preceding sections set out our assessment of their success in achieving the goals of the Superfund program.

What *general* lessons can be derived from the case studies? This chapter concludes with a set of tentative conclusions regarding what works, what might work, and what does not seem to work in the Superfund program.

At a basic level, all strategies "work" in that they eventually bring about some form of cleanup. But each strategy seeks to maximize one objective of the program, to the potential detriment of the others. Thus, each approach can lay claim to success on one dimen-

sion, but each is also subject to criticism for failure in comparison to its alternatives.

This first point should be clear from the analysis in this chapter. The prosecution strategy emphasizes the importance of minimizing government costs in cleanups. The accommodation approach seeks to reduce transaction costs. The public works strategy attempts to maximize speed of cleanup. It is not clear to us that any of the three is any more likely to result in a "cleaner" clean. But at any site in moderately serious condition, it is seldom possible to have a speedy cleanup, with minimal transaction costs, at no cost to the government—at least if there are technical limitations on what constitutes an effective remedy.

In addition to delivering on differing combinations of program goals, strategic choices also generate what might be termed *characteristic errors*. Practitioners of each strategy, when confronted by PRP resistance, may overreact in different ways. Use of a prosecution strategy may encourage resort to coercive tools, such as premature unilateral orders or threats, that may either create subsequent problems or provoke further resistance. Practitioners of accommodation may be willing to concede too much in an effort to establish rapport. Those employing a public works approach may resort too quickly to the use of removal powers when further negotiations are indicated.

Choosing a strategy thus involves setting priorities among competing program objectives. While these trade-offs may be discussed in broad policy terms in Washington, our experience suggests that such discussions occur rarely, if at all, at the level where implementation decisions actually are being made in individual cases.[33] A useful first step toward rationalizing the implementation of Superfund would be for regional EPA staff to decide what they want to achieve at a particular site. In most cases, it appears that the implicit goal is to achieve everything—a fast, cost-free, fully effective cleanup with no transaction costs. Which objectives are ultimately obtained at the end of the day, and which are sacrificed, frequently seems more a matter of chance than of conscious design.

All Superfund sites require governmental effort and the assumption of some of the costs associated with cleanup. Despite punitive rhetoric, we do not believe the experience of Superfund

gov't will inevitably have to pay some of clean up of (burden sharing)

suggests that all the costs and risks associated with cleanups will be borne by private parties.

The expenditure of public funds is necessary at sites where no private responsible parties can be located, or where they lack the financial resources necessary to effect a cleanup. But even at sites where deep-pocket PRPs are present, the government inevitably assumes some costs. These administrative and cleanup costs may be covered at the outset and subsequently forgiven in implicit mixed-funding arrangements. Some costs may take the form of future administrative expenses that are not assessed against the responsible parties. Some costs may be covered directly by means of a governmental share in an explicit mixed-funding agreement. Some costs may simply prove uncollectable. A 1991 report by the General Accounting Office indicated that in 1989, the EPA recovered all of its costs only 20 percent of the time.[34] As indicated above, these costs accrue to the government because of a variety of factors: the dynamics of settlement procedures before federal judges; a felt need to accommodate the cooperating parties in a PRP-led cleanup; an effort to address the problem of "orphan shares" whose burden would otherwise fall entirely on the shoulders of cooperating PRPs.

It is our judgment that a government strategy to give no quarter, to apply the broadest conception of joint and several liability so as to place the entire cleanup burden on the PRPs, is unlikely to meet with success. Indeed, our cases suggest that it may well be counterproductive. Failure to conduct an adequate PRP search at the outset, for example, only shifts administrative costs to a later point in the process, when positions have hardened and the investigation has taken on an adversarial, legalistic tone, with far greater costs the likely result for all concerned.

Acknowledgment that, despite joint and several liability, the government will bear part of the burden in cleaning up Superfund sites, may actually prove liberating. Once it is recognized that some form of burden-sharing frequently is inevitable, even desirable, the inquiry can move on to determination of an appropriate allocation of that burden. The financial, political, and legal imperatives of the Superfund program all point toward PRP assumption of most of the costs of cleanup. But recognition that no moral imperative *compels* a 100-percent solution, and that such a complete government victory is unlikely in any event, may allow creative settlements to be negotiated at an earlier stage of the process, rather than as last minute concessions to the pressure of judges, local residents, or "bean counters" in EPA headquarters. This assertion gives rise to a related observation:

The government and the private parties involved with Superfund sites typically are concerned with minimizing different kinds of costs. The potential therefore exists for trade-offs that will result in net benefits to both sides.

There are clear limits in the extent to which the government will share the costs of remedial action at Superfund sites where PRPs possess the requisite financial resources. Superfund revenues may be used to cover some of the costs of orphan shares, especially when the potential exists for the government to collect these expenditures from nonsettling PRPs. However, political and economic imperatives limit the extent of such participation.

Our discussions with PRPs and their attorneys suggest that while actual cleanup costs are matters of substantial concern to them, the risk inherent in the almost unlimited *future* liability at Superfund sites is frequently a more serious impediment to settlement. Government attorneys usually insist that the consent decrees signed by major responsible parties include broad reopener clauses which allow the government to reinstitute legal proceedings in the event the remedy specified in the ROD proves to be ineffective, or if any unforeseen circumstances develop at the site that suggest the need for additional remedial measures. Indeed, the refusal of government attorneys to allow the government to accept *any* element of future risk—even in regard to the government's own expenses in supervising a remedy—characterized two of the complex cases we examined (Tybouts Corner and Laskin Poplar). One of the major incentives for a defendant to settle a court case, the promise of a final end to the potential losses involved in the action, thus is missing in Superfund litigation.

De minimis parties are willing to pay a premium on top of what would otherwise be their proportionate share of a remedy, in order to buy their way out of future liability. This fact indicates that *de maximus* PRPs also may be willing to absorb substantially higher immediate costs if they can escape from the imponderable future burdens that accompany Superfund settlements.

The Superfund is limited, and in light of the current scarcity of public funds, is unlikely to be sustained by revenues other than those accruing from the dedicated tax on chemicals. Further, CERCLA and SARA make explicit that the burdens of site cleanup are to be borne primarily by the private parties made liable by these statutes. That those private parties might be willing to pay a premium to buy their way out of future liability at a site suggests the possibility of "*de maximus* buyouts*": PRPs might be offered the possibility of paying *more* than their share of predicted cleanup costs in exchange for a release from future liability. These premiums could then stay in the fund as protection against the occasional failed remedy. While such a method might not work in all cases, it could provide a powerful inducement to PRPs to settle their cases and clean up their sites.

Just as different implementation strategies promote different general notions of program success, so each may be more or less appropriate to the situation at a particular site.

Our case studies suggest that factors unique to particular sites limit the possibility of achieving the divergent goals of the Superfund program. At sites presenting large or unusually complex technical problems, speedy remediation may be difficult to achieve. At sites dominated by uncooperative PRPs, high transaction costs may be unavoidable. At sites where no deep-pocket PRPs are present, or where evidentiary or other problems weaken the government's case, a substantial part of the cleanup costs may have to come from public funds. In light of these observations, we offer the commonsensical proposition that enforcement strategy should be matched to site characteristics.

We would be reluctant to put forward so seemingly obvious a proposition had it not been so frequently ignored in the cases we studied. Instead of a discerning choice from the broad range of statutory tools and implementation strategies offered in Superfund, regional decisionmakers were guided either by the current policy being espoused in EPA headquarters, or—more frequently—by the limited menu of techniques they and their staff appeared to know how to use. Superfund sites undoubtedly differ in systematic ways from region to region, but these tendencies cannot explain the strong regional preferences toward one or another of the implementation strategies that we have observed, preferences that often seemed to be applied in an unthinking manner to all sites in a region, regardless of their applicability to individual circumstances.[35] At least a part of the EPA's difficulties in Superfund may be attributable to the "hammer and nail pathology": if you only know how to use a hammer, every problem looks like a nail.

We conclude that the question, What works?, with which this research began, might be more appropriately phrased as, What works best, in which circumstances? We obviously are not in a position to provide definitive answers to this question after the study of six cleanups in three regions. It would take a far more ambitious study to take account of the many sources of variation across sites that should probably influence choice of implementation strategy: size, technological difficulty, and projected cost of the cleanup; the feasibility of various remedial alternatives; the number, financial resources, and cooperativeness of PRPs associated with the site; the strength of the evidence against the responsible parties; the stance taken by state officials. However, we do believe that our case studies suggest some tentative conclusions about the site-specific conditions that should influence choice of implementation approach.

While many variables should be taken into account in this decision, we focus attention on four: (1) predicted cleanup costs, (2) technological complexity, (3) presence or absence of deep pocket PRPs and their degree of cooperativeness, and (4) the need to expedite cleanup.

A *public works* approach to cleanup is most appropriate in sites characterized by:

—Low projected cleanup costs

—Few technical problems, especially where off-site disposal is both appropriate and feasible

—The absence of cooperative or motivated deep-pocket PRPs

—A pressing need for speedy remediation

A *prosecution* approach is most applicable in sites with:

—Moderate to high projected cleanup costs

—Deep-pocket PRPs, all of whom are uncooperative

—No special need for speedy remediation

An *accommodation* approach is most appropriate in sites with:

—Moderate to high projected cleanup costs

—One or more cooperative, deep-pocket PRPs

—No need for immediate remediation

The choice of a public works approach is constrained more by the nature of the site than by the configuration of PRPs. Legal and financial imperatives constrain even the most aggressive notion of what can constitute a removal. Even at a relatively small site, however, the presence of cooperative PRPs with sufficient financial resources to do an effective cleanup would suggest a PRP-led cleanup consistent with the accommodation approach.

At larger, more complex sites, the real choice is between prosecution and accommodation. Here, we believe that the major determinant of approach should be the potential for PRP cooperation or noncompliance. Kagan and Scholz's general analysis of regulatory strategy suggests the choice facing the EPA at individual Superfund sites:

> To treat every firm as an amoral calculator, whereby any deviation from specific regulatory rules is met by legal penalties, burdens the economy with unnecessary costs. It also breeds legal and political opposition on the part of good corporate citizens who are offended by being forced to meet unreasonable requirements and by the perceived injustice of punishment pursuant to legalistic rule application. Conversely, if regulators were always to act as responsible politicians or consultants, and always withhold penalties in hopes of convincing or teaching the company to do "the right thing," the amoral calculators will take advantage of their flexibility. The relevant question, therefore, both from the standpoint of explanatory or predictive theory and from the standpoint of regulatory strategy, is not which theory to use, but when each is likely to be appropriate.[36]

This key question raised by Kagan and Scholz may be easier to answer in the context of Superfund than in that of regulatory programs. Most such programs face the regulated population on a wholesale rather than a retail basis; it is therefore difficult to tailor an agency's response to any individual corporation. By contrast, Superfund involves intense interaction between the agency and a small group of businesses over an extended period. Further, the set of PRPs at Superfund sites frequently includes some of the "Usual Suspects": the Fortune 500 corporations whose industrial activities have made them PRPs at a number of sites nationally. There is thus the potential (seldom realized, we suspect) of building some institutional memory on the question of which PRPs can be expected to be the "responsible citizens," and which the "amoral calculators."

The potential benefits of a choice of implementation strategy that takes into consideration the context of the site, and the agency's priorities for it, would rationalize and improve program performance. At a minimum, it would give policymakers concerned with Superfund a fair measure of whether this extraordinary piece of environmental legislation is workable, or if it needs to be replaced with a more traditional regulatory or public works scheme.

Seven

Ruminations on Reform

SUPERFUND HAS been roundly criticized for lengthy delays, high costs, and limited accomplishments. Some of these criticisms have ignored the program's successes—most notably the substantial accomplishments of the emergency removal program. Others have discounted the start-up time needed to construct the bureaucratic, scientific, and legal apparatus for a program of the complexity and novelty of Superfund. Further, in condemning the lack of progress at individual sites, critics have not always taken full account of the time required to resolve scientific and engineering imponderables inherent in the attempt to clean up a major hazardous waste site, or to convince an unhappy chief executive officer that her company must pay substantial sums of money to clean up a site for which it bears little direct responsibility, or to litigate against other PRPs who refuse to settle voluntarily.

Nevertheless, we suspect that the overall picture that emerges from our descriptions of the operation of Superfund on the ground, at actual hazardous waste sites, will do little to allay concerns in the environmental policy community about either the operation or the design and organization of America's hazardous waste cleanup program. While some of the sites we have described moved toward cleanup more expeditiously than others, with less acrimony and fewer seemingly gratuitous transaction costs, the overall picture presented is one of lengthy delays, high costs, and conflict within the Environmental Protection Agency, between the EPA and other government agencies, between the government and the potentially responsible parties, and among the PRPs themselves. It is not easy to assess which of these observed pathologies arose because of faulty implementation, and which were based on more fundamental deficiencies in program design and agency organization.

Whether Superfund is assessed in terms of an aggregate balancing of money spent with cleanups accomplished, or with respect to the events occurring at typical hazardous waste sites, common sense suggests that improvements are in order. The recommendations that conclude the previous chapter are based on empirical observations of the strengths and weaknesses observed in the various strategies adopted by the EPA to implement Superfund. Recognizing that none of them is especially profound, we conceive our suggestions to be analogous to the home builder's advice to "put the bathroom closer to the bedroom."[1]

The limited scope of our recommendations is related to the original purpose of this research. As we have emphasized from the outset, our focus was on policy implementation and the strategies used by government to put enacted programs into operation. Thus, throughout this project, we took the underlying statutory and organizational structure of the Superfund program as a given, grounded in the legal, resource, and bureaucratic imperatives to which the EPA must respond. Our initial focus on regions and their activities contributed to this orientation: regional officials are in a position to interpret administrative and statutory directives, and to select among various implementation options, but they are constrained by the materials available and the choices permitted. Put another way, we have evaluated to this point how the EPA has used the tools at its disposal. But there are structural or macro-level considerations that also contribute to the observed outcomes of the Superfund program.

Policymakers in Congress, the White House, and the EPA are in a position to make more fundamental changes, to alter the givens, and change the legal and organizational constraints within which the Superfund program operates at individual sites. We have not discussed the desirability of altering these elements of Superfund because the case selection methodology that underlay our study did not permit a balanced overall picture of the program. And, because all the cases studied were handled within the existing statutory and administrative framework, no systematic evaluation of alternative program designs was possible.

While we have no empirical data that measures the current performance of Superfund against alternative legal and organizational arrangements, neither, we suspect, do other researchers. If this study is to be of use to those contemplating fundamental change in the Superfund program, it should at least address these concerns, however tentatively. Therefore, we temporarily (and perhaps ill-advisedly) divest ourselves of scholarly caution in this final chapter and contemplate the implications of what we have learned for broader issues of Superfund's basic design and the organizational structure within which it is implemented. We close the chapter, and the book, with even more speculative homilies on the lessons of Superfund for the increasingly privatized government programs of the present and the foreseeable future.

In the next two sections of this chapter, we examine broad aspects of Superfund that we have previously taken as givens. The first concerns the organization of the program, and provides a look at a series of issues associated with the EPA's responsibility for obtaining the cooperation of the diverse participants and interests at individual sites. These problems are all related to what policy analysts call the complexity of joint action. In this section we also discuss related problems caused by policy initiatives that substitute process for agreement—what is termed in the social-work community an advocacy framework for making decisions—and the perverse individual and organizational incentives that frequently tip the scales against agreement in such contexts.

The second section concerns Superfund's liability scheme—particularly joint and several liability—which tends to accentuate the problems of obtaining agreement, even as it simultaneously drives the settlement process. We raise the issue of joint and several liability with some trepidation, since it is probably the most controversial aspect of the Superfund program. Yet most current proposals that contemplate major change in hazardous waste cleanup policy involve alteration or elimination of Superfund's reliance on strict, joint and several liability. We believe that our case studies—while hardly determinative of the complex policy issues involved—can illuminate aspects of the liability scheme that may have escaped previous consideration.

This discussion obviously does not exhaust all the serious alternatives for changes in the statutory design of Superfund, or its organizational structure. We discuss only those issues of potential reform and change upon which our research casts some light.

The Complexity of Joint Action

The three Superfund implementation strategies we described in preceding chapters—accommodation, prosecution, and public works—share two characteristics: each relies on the cooperation of multiple actors with differing perspectives and goals, and each involves transferral by the EPA of some (or all) of the burdens associated with Superfund cleanups to others. No matter how a specific strategy suggests approaching a major Superfund site, the resulting cleanup inevitably will involve an enormous amount of raw technical work, administrative attention, and numerous judgment calls on a daunting variety of legal, scientific, and political questions. It will also require substantial amounts of money, obtained primarily from the private businesses (and, increasingly, local government entities) that find themselves labeled potentially responsible parties. These elements are the inevitable baggage of a program explicitly designed to place most of the burden of hazardous waste cleanup on the parties held to be responsible for making the mess in the first place.

Prosecutorial strategies depend on liability doctrines and administrative powers to compel PRPs to assume cleanup responsibilities. Accommodation strategies focus on stressing mutual interests in facilitating a cleanup, but only in the context of persuading PRPs to assume as much of the burden as possible. The public works approach, while envisioning a more active role for the EPA in particular parts of the process, must still depend on the PRPs ultimately assuming responsibility for cleanup costs if the cleanup program is to remain viable politically, administratively, and fiscally.

Besides PRPs and Superfund program personnel, essential players include state agencies (by means of their decisions on applicable or relevant and appropriate standards, or ARARs, and their political position in a federal system), other EPA programs (such as Clean Air, Clean Water, and the Resource Conservation and Recovery Act, or RCRA), other federal agencies (the Coast Guard and the Fish and Wildlife Service, for example), the Department of Justice, and federal judges. The cast of characters and the preferences each expresses depend on the issues raised at a specific site. This situation produces multiple difficulties, since interorganizational relationships must be continuously redefined.

Because each actor or set of actors commands some essential resource (money, authority to sign off, technical competence), these players often are in a position to deny—through obstinance or timidity, or because of rigidity in their procedures or the requirements of their bargaining positions—what is required for successful settlement. Each views the world in the context of a distinct set of individual and organizational concerns that can impede prompt site remediation. The problems arising from this situation provide a definitional example of the complexity of joint action.[2]

Government programs characterized by such complexity are necessarily prone to "implementation games" which waste time in conflicts, divert money from programmatic purposes, or displace goals.[3] As the case studies reported in this book show, the process of internal clearance of case-specific decisions in Superfund widely disperses the power to say no. Other government agencies take a regulatory role in their relations with the EPA in Superfund cases: they impose requirements, make judgments on the basis of information gathered by those seeking permission to go forward, and assess the desirability of any action in terms of its consistency with their own regulations and organizational objectives.

Because of these circumstances, the current system is inherently risk-averse and stacked against errors of commission. It is structured so that obstruction is cheap, while facilitation is labor-intensive. Authority is divided among participants who must assume full responsibility for compliance with a given area of regulation or the maintenance of a congressional or state-mandated requirement. Thus, while most of the credit for a successful cleanup or settlement accrues to the EPA, many of the specific risks involved in the cleanup must be assumed by those who granted the necessary permissions.

Perhaps the clearest organizational manifestations in Superfund of the complexity of joint action are associated with getting approvals from government agencies outside the Superfund program. Many approvals involve technical requirements that define the standards for evaluating cleanup alternatives: compliance with RCRA, compliance with the Clean Water Act, compliance with state ARARs.

As a practical matter, the clearance process typically requires sign-offs from a variety of agencies within the EPA and in other parts of the federal bureaucracy. In addition, site-specific negotiations frequently require dealing with state governments on issues ranging from sign-offs on state and local environmental requirements, to intergovernmental conflicts over which standards are applicable, to disputes over the extent to which the states should contribute matching funds or take on risks as the final custodians of sites. And, of course, there are the relationships between the EPA and the Department of Justice which emerge because the Justice Department serves as the EPA's legal counsel in court actions. Like other agents, the Justice Department has considerable latitude in defining the requirements of Superfund settlements and consent decrees, and from all appearances it exercises that discretion freely.[4]

Statutory and administrative requirements thus produce high governmental transaction costs that result from the need to obtain agreement and cooperation from a wide range of governmental actors responding to diverse and sometimes conflicting incentives. While some of the machinery for dealing with this issue has become routinized—such as the process of referring Superfund cases to the Justice Department—other parts have to be assembled anew for each case. This system of clearances increases the cost of negotiating the implementation maze and results in fewer rats running a slower race. Two further problems thus arise.

First, those whose responsibility it is to obtain approvals find themselves dealing with organizations constituted to do something other than facilitate Superfund settlements. Often, as is the case with the Department of Justice, organizational incentives press in other directions. Or, as is the case with RCRA, Superfund issues have only recently begun to emerge and procedures have not yet been fully articulated for dealing with program disjunctures. Finally, as other organizations are asked to sign off formally on Superfund agreements, they bear the risks associated with potentially controversial or incorrect decisions, and the danger of being seen as letting up on their mission. Such a calculus is stacked against pro forma cooperation if judgments must be made instance by instance.

Second, PRPs and the outside world receive the impression that government is not acting in an understandable and consistent fashion. Not presenting a unified face to the world, in turn, produces other problems in the EPA's negotiations with PRPs. As indicated previously, PRPs settle largely because they find the certainties of settlement preferable to the uncertainties of failing to agree. But the process we have described above generates considerable uncertainty and costs of its own. It is often time-consuming; it makes cleanup standards a moving target by revising and reopening decisions as the process progresses; it creates uncertainty because critical permissions are withheld until the end of the process—as records of decision are written and administrative orders issued—or until the remedial design and remedial action (RD/RA) phase, a time when PRPs may be deciding whether or not to settle.

Defining Program Goals in an Advocacy Framework

Superfund is especially susceptible to the complexity of joint action because of an issue raised in chapter 2 and again in chapter 6: the imprecise nature of the goals of the Superfund program, both at the level of aggregate program performance and at individual sites. Superfund is precise about who should pay and who should be consulted in determining what must be done. The message concerning *what* should be done is more confused because the requirement is operationalized as a process rather than as a performance standard. How are program managers to know if a remedy is clean enough? If the appropriate parties have been consulted and the procedural steps have been followed.

Lawmakers probably finessed the critical issue of "how clean is clean?" because they were unable to formulate precise cleanup standards across sites with many different problems and characteristics, or because they could not agree on what those standards should be, or because they were unwilling to establish priorities among widely held and frequently competing values. At the same time, they were unwilling to leave decisions wholly to negotiations between PRPs and the EPA. So it was both easier and more attractive to determine who should be consulted and what should be considered rather than what should be done.

Defining a process rather than a set of standards was an effective way for policymakers to satisfy many political objectives at once. The need for widespread support in legislative coalition-building could be satisfied by creating a long list of relevant considerations in defining "how clean is clean": a preference for permanent remedies, a requirement that "no action" be considered as an alternative at every site, an expectation that some form of cost-benefit calculation be performed, the need to consider all relevant environmental standards found in federal or state regulations. Freed of the requirement of choosing among or balancing these competing alternatives, policymakers simply could add to the list, depending on whose support was needed. The result was a program in which no important consideration was left out.

But the lack of discrimination in the list of cleanup considerations posed another problem. A requirement that everything be considered, that everything be important—in a context in which trade-offs among such concerns are inevitable—gives the implementing agency substantial discretion in defining what must be done. If Congress had confidence in the judgment, competence, and independence of the implementing agency, this risk might be acceptable. But relations between Congress and the EPA historically have been rocky. Mistrust of the EPA repeatedly has led Congress to define agency discretion more narrowly than the EPA has wanted.[5]

Because Congress distrusted the implementing agency, it was unwilling to allow that agency to strike a balance among competing values. What to do? The solution in Superfund, as in a number of other policy areas, has been to create a process centered on what we term an *advocacy framework*.[6] Consideration of important values was institutionalized in the form of participation requirements.[7] "Every interest was to have its watchdog."[8] As a result, the legislation contains explicit and implicit requirements that various groups—each concerned with a particular value—participate in decisionmaking. Each would have some power to block action as leverage to ensure consideration of its own policy agenda.

The involvement of responsible parties would ensure that the costs of remedies be considered. Public participation by local citizens and national environmental groups suggests that more stringent standards will be considered. The involvement of other federal and state regulators—through the ARARs process—ensures that the professional guardians of other environmental standards can see that these standards are observed. In this process, the test of a good cleanup becomes the capacity of diverse participants to agree on a single standard. Any remedy that gains the consent of so many different parties must—by definition—be a good one.

The locus of this advocacy framework system is remedy selection, at the point when the record of decision is finalized, as well as at the point when consent decrees formalize agreements between the federal and state governments and the PRPs on what is to be done and who is to pay for it. This focus on formal decision points has three consequences.

First, these decision points constitute almost the only formal opportunities for expanded participation; most other parts of the remedial process involve only the EPA, the responsible parties, and their agents. As a result, other actors have strong incentives to push their case at these specific times and to formalize arrangements in as much detail as possible. More continuous participants might prefer looser formal agreements that they can modify if circumstances change. Not surprisingly, the load at these key points creates bottlenecks that contribute to delay.

Second, the mode of discourse at these critical decision points is to advance, in a narrowly focused fashion, the special concerns of each participant. Such a framework encourages actors to push their particular concerns in the expectation that breadth of participation will take care of the larger range of other considerations. Breadth of participation, in short, frees each participant to take responsibility for only a part, thereby simplifying his or her tasks considerably. Not surprisingly, when a final result is achieved, it is often not the one desired by any single party; each may be dissatisfied with some part of what is ultimately agreed to.

Third, the phases of cleanup with the potential for more cooperative relationships, such as those dealing with technical and engineering problems, will occur before and after the more participatory parts of the process. The range of groups engaged in setting cleanup standards would, in such a system, be the largest and most diverse. The range of groups engaged in the less conflictual aspects would be narrower and drawn from those most familiar with a given site and with each other.

While this advocacy framework process is not as inherently adversarial as the judicial process, it nevertheless sets interests against one another. Almost by definition, the decisions to be made were too technically difficult or controversial for Congress to resolve in the authorizing legislation. Not surprisingly, this controversy extends to the surrogate process created to make those same choices at individual sites.

Thus, some of the observed difficulties of conflict, delay, and dissatisfaction that rise out of the complexity of joint action can be attributed to the basic decision to define cleanup standards by an advocacy framework. The framework encourages many participants to define their interests in mutually exclusive terms; each focuses single-mindedly on his own set of values and relegates guardianship of other values to other participants. Conflict results because decisions require that priorities be established among these values. Delay follows because the absolute numbers of participants and interests can be quite large, and accommodating individual concerns (even when they are not mutually exclusive) takes time. In such circumstances, some degree of dissatisfaction with decisions is inevitable, and recourse to the courts is always a possibility for those most aggrieved.

If substantial impediments to the effectiveness of the Superfund program caused by the complexity of joint action and the related advocacy framework are to be reduced, and if those with responsibility for specific cleanups are to make more informed and creative use of the policy devices in their Superfund toolbox, the agency must consider how to reconfigure the distribution of discretion within the agency, and between the agency and other relevant actors. The abandonment of an advocacy framework is unlikely, and perhaps even undesirable. But some of the observed pathologies brought about by this approach to remedial decisionmaking can be reduced.

Three potential approaches to these problems recommend themselves. The first is creation of conditions favorable to greater participation by "fixers," policy actors operating in the highly specific context of particular cases, working to coordinate action among diverse individuals and organizations, and to overcome obstacles.[9] As is shown in this chapter, this path calls for a "bottom-up" or "street level bureaucrat" approach.[10] The second approach is to move in another direction altogether: toward the creation of a more comprehensive traditional government agency that encompasses a larger number of essential functions within its own organizational boundaries. While such a body might provide greater discretion to those at its bottom, the main thrust really implies a "top-down" approach.[11] Finally, some means could be found to realign the incentive structures of the various organizational participants in Superfund settlements to recalibrate the balances that are currently weighted against agreement.

A Bottom-Up or "Fixer" Solution

Fixers require informational resources and a degree of power; when they function effectively, they facilitate cooperation between disparate actors.[12] Most governmental actors involved in Superfund activities do their jobs almost entirely within their own organizational structures, whether as upper-level EPA officials, state and federal agents responsible for pressing particular standards, or lawyers employed by the Justice Department. It is often the case that "what you see depends on where you sit."

Some government officials do have a wider network of information sources concerning the varied preferences of both governmental and nongovernmental actors. The remedial project managers (RPMs) and assistant regional counsels—those who negotiate with the PRPs, deal with contractors and consultants, and serve as the contact points for state-federal and EPA-Justice Department discussions—are well-situated to identify bottlenecks in the process. These officials frequently have information on what is impeding settlement and are good candidates for the fixer role.

While they have access to information from the varied sets of actors, RPMs and assistant regional counsels are in a difficult position because they have little capacity to break the jams and few alternatives for going around them. Not surprisingly, they are often frustrated when clearances stall in other parts of the EPA or at the Justice Department, or when they must reconcile competing substantive and procedural standards, or unravel tangles that threaten to delay or abort negotiations or the formal process of decision.

The difficulty is that the present structure is both organizationally and normatively biased against making the kind of concessions that are frequently necessary to achieve settlements in Superfund cases. The system provides frontline personnel with little discretion to assume any of the costs and risks involved in Superfund cleanups. If these individuals are to play a fixer role, they need additional discretion over the subjects of negotiation and some means of ensuring the cooperation of other governmental actors in the process.

While it might make sense to grant greater discretion and power to RPMs, their supervisors, and the other regional personnel directly engaged in site negotiations, this prospect seems unlikely for several reasons. First, the EPA has demonstrated little confidence in its line personnel, perhaps because of their high rate of turnover. Second, Congress—and the states, through the congressionally mandated requirement to meet relevant state requirements—has not chosen to legitimate negotiations on cleanup standards. While some negotiation over the apparently nonnegotiable is inevitable, the present structure requires a complex system of sign-offs in order to spread the responsibility for a decision to a point where it is bureaucratically tolerable. And, in the case of prosecutorial strategies, the Department of Justice is unlikely to relinquish its preeminent role in controlling cases in litigation and assessing proposed concessions in terms of broad questions of precedent.

An alternative to using fixers within the agency is reliance on outside parties, "honest brokers" who can bring disparate parties together and identify areas of mutual advantage. Environmental policy provides many examples of the successful use of alternative dispute resolution techniques in a variety of areas. The most prevalent use of ADR in Superfund has been in helping PRPs reach decisions on allocating Superfund liability. The EPA and Justice Department have resisted the use of outside mediators in disputes between the government and the PRPs. There is not, however, any theoretical impediment to such efforts, and independent third parties could very well increase the confidence and trust of participants in a process in which these commodities are frequently in short supply.[13]

A Top-Down Solution: Traditional Government Organization

The second approach to the problems inherent in the complexity of joint action is building a more comprehensive government organization, thereby augmenting the EPA's current structure with increased capacity and wider responsibilities. Scholars recently have rediscovered the virtues of traditional governmental organizations. They hold that such organizations often have high rates of success.

> What is more, these successes are related to the inherent characteristics of this tool—the fact that it internalizes transactions which indirect government must negotiate across institutional boundaries, maintains solid in-house expertise, minimizes legalism, and provides a more stable framework for bargaining with external interests.[14]

While it is unlikely that the EPA can be made into the comprehensive agency described in the above passage, there is room for movement in that direction in Superfund activities. There are, first of all, some areas where building a greater capacity within the EPA—without reliance on outside consultants and others to do its work—would not involve major alterations in organization, procedures, or expenditures. The agency might begin by obtaining more comprehensive information on PRP liability at Superfund sites. Our case studies of Tybouts Corner and Laskin Poplar make clear the folly of using litigation—particularly third-party litigation—to develop this information. Such a process is slow and extraordinarily expensive (although its cost, primarily in the form of legal fees, is borne mainly by the PRPs). The EPA has in fact increasingly relied on in-house investigators to collect information on PRP liability at waste sites. Ultimately, the PRPs could be made to pay the costs of such activity; this may seem like a bargain compared to development of information through adversarial procedures.

A related area of reform is the generation of information about the amounts and types of waste contributed to a site by various parties. Our case studies suggest that the practice of generating waste-in lists in Region IV, and the limited use of informal nonbinding allocations of responsibility in Region V, accelerated settlement negotiations and reduced aggregate transaction costs. The EPA already has legal authority to get the needed information—authority the PRPs can only obtain for themselves through court action. What the agency has lacked is the will or the personnel to do the requisite work.

Another area in which the EPA's internal organization might be recalibrated is in the creation of an institutional memory concerning individual settlement discussions. Negotiations at Superfund sites typically take a long time. EPA turnover at the bottom is high, and lateral moves by top and middle managers are common. As a result, few major sites engage government personnel whose experience spans the whole process. What the government side knows about a given site seems in large measure to be determined by the paper record in the files, or by the passing of information orally from one generation of RPMs and lawyers to the next. By contrast, the principal agents of the PRPs engaged in negotiations and cleanup activities at a given site are usually a much more stable group.

It is difficult for government officials and PRP representatives to build relationships of trust when EPA personnel and other government actors come and go so frequently. Indeed, a common complaint among PRP attorneys is that understanding and goodwill built up with one set of government officials are seldom transferred to their successors. This problem is compounded as control over cases moves from the EPA to the Justice Department, to different sets of people and organizational cultures. It would make sense for the EPA to acknowledge that its personnel problems have a practical impact on what the agency seeks to build—a cooperative relationship with PRPs in an accommodation mode, or at least a predictable relationship in a prosecutorial mode—and to make adjustments accordingly.

Unfortunately, the present system standardizes case-handling procedures only to the extent that government behavior is shaped by laws, regulations, and various organizational procedures. A modest investment in institutional memory—comprehensive site histories (rather than brief summaries generated to meet formal requirements), regular briefings and debriefings, the creation of transition teams to manage changes in personnel—would help provide both the EPA and the PRPs a clearer record of cooperation and conflict on which to build their long-term efforts.

Another means of addressing the problems associated with the complexity of joint action would be for the EPA—in consultation with the other governmental actors—to develop routine general standards and agreed-upon areas of discretion in dealings with agencies both within and outside the EPA. One illustration of such an effort can be found in the attempt to produce a model consent decree.[15] Another could involve setting uniform national standards for the various ARARs applicable to similar groups of Superfund cases.[16] These issues presently are negotiated ad hoc with a host of government agencies for each cleanup. A move toward some uniformity in ARARs and consent decrees could eliminate the necessity to reinvent the wheel for each cleanup, and could also allow Superfund personnel to assume both the risks of interpreting and applying these standards and the authority associated with more comprehensive control of the process.

A final improvement would entail reducing the number of critical actors involved in decisions at individual sites by consolidating some of the required sign-offs or allowing the EPA to serve as custodian of some of the interests currently protected by other agencies. In light of the advocacy framework that underlies the current Superfund program, this is probably not an area for broad reforms as much as one in which fine-tuning and adjustment are required. Federalism and agency structures put constraints on just how much can be internalized within the EPA. Still, we suspect that modest changes could reap substantial improvement in Superfund decisionmaking.

Redistribution of authority relationships based on the tasks at hand could place more authority in the hands of those with the statutory responsibility for cleaning up Superfund sites, rather than in organizations whose priorities are unrelated to facilitating cleanups. This endeavor might be easiest to accomplish in relationships between the Superfund program and other programs within the EPA concerning ARARs and other potential stumbling blocks to settlement. In this context, we note that the EPA has created an administrative position with formal responsibility over the entire Superfund program. That creation of this position came only after ten years of program operation may strike some as puzzling, but it certainly helps move the EPA in the direction we are suggesting.

One area in which ambiguous grants of authority have proved particularly troublesome is in the relationship between the EPA and the Department of Justice in site-specific settlement negotiations. The EPA considers itself the client of the Department of Justice. Yet the Justice Department, as one of its high-ranking officials told us, "has as its client the United States of America." The ambiguity over who is in charge, and who will bear final responsibility for the risks and potential benefits inherent in any set of settlement negotiations, is a source of conflict in many Superfund cases. We already have indicated that the priorities of the EPA and the Justice Department may diverge: the former is frequently most interested in obtaining speedy resolution of a case and moving on to the cleanup; the latter is often more concerned with not surrendering legal ground in settlement discussions, or with avoiding the establishment of precedents unfavorable to the perceived long-term legal interests of the government.[17]

The practical compromise apparently reached between the EPA and the Justice Department is to give the department the lead role in negotiations and an effective veto over any proposed settlement once a case is potentially in litigation. This control determines which legal positions are maintained and which settlement provisions are offered and accepted. While the nature of control varies—it seemed to have been stronger in the Tybouts Corner case in Region III than in the Laskin Poplar case in Region V—it is nevertheless present across cases.

We suggest that a somewhat different division of labor might be considered, one in which the Justice Department maintains control over potential precedent-setting legal concessions and maneuvers, but relinquishes control over the site-specific bargaining to those more closely tied to the Superfund program. Some movement in this direction has already occurred within particular regions. So, for example, the assistant regional counsels in Region III (and, to some extent, in Region IV) consider themselves to be lawyers for the program staff who serve as their clients. In Region V, however, the lawyers have the upper hand, and the program staff must work with the ever-present idea that they will ultimately have to help defend their actions in litigation. We think that the more informal lawyer-client relationship established in Region III is a model of how internal authority might best be deployed to place authority and responsibility for program success in the same hands.

Modifying Incentives

The preceding discussion was based on an assumed linkage between settlements at individual sites and organizational relationships among the Superfund program staff, and between the Superfund program and the various organizations that must coordinate their activities to achieve an agreement on cleaning up a site. In this section, we look at less formal approaches to the problems caused by the complexity of joint action, and ask if the modification of incentives, independent of any proposed structural changes, might promote a more effective and efficient cleanup program. A common theme in analysis of the case studies has been the perversity of many of these incentives, and the resulting organizational and individual resistance to obtaining closure in site negotiations.

While it is reasonably clear that Superfund program staff have the greatest personal and organizational interest in reaching satisfactory agreements and cleaning up toxic waste sites, even the RPM and his or her immediate supervisors operate with mixed incentives. Their broad interest in cleaning up sites is overlaid by other objectives: maximizing the number of countable events (or "beans") that serve as the

EPA's indicators of personal and regional success in Superfund, avoiding political flack caused by appearing too friendly with PRPs, and making no obviously incorrect decisions concerning complex technical issues involving great uncertainty. Depending on circumstances, each of these concerns—particularly the last—can work against acceptance of almost any potential agreement with PRPs. Thus, we suggest that the structure of incentives within the Superfund program itself be reassessed, with greater weight given to rewards for substantive achievement rather than simply for meeting procedural benchmarks.[18] The goal should be to target incentives to desired behavior, and to reduce the chances that these incentives will encourage perverse behavior.

Several strategies recommend themselves. Perhaps the best starting point would be the elimination of perverse incentives: those that encourage dysfunctional actions and inactions. The "bean count" system of quarterly performance assessment on a limited set of indicators provides the clearest example. While it is true that Congress defined a number of performance measures and numerical benchmarks by which to assess aggregate program performance, this set of statutorily based measures need not constitute the only, or even the most important, dimensions of program success within the EPA. It certainly need not constitute the primary criterion by which additional resources are distributed among the regional offices.[19] For example, rewarding a region for the number of enforcement actions it refers to the Department of Justice may only serve to increase the number of slipshod, ill-prepared cases dumped at the door of the Justice Department. (We have, in fact, heard complaints that some regions engage in this form of behavior.)

Several reforms might be instituted to mitigate the perversity of the incentives that arise from the EPA's system of Superfund performance measures. These include: longer reporting periods (quarterly reports give the illusion of greater control while diminishing incentives for greater quality); a wider-ranging, more carefully designed set of indicators based on achievement of broad program objectives, rather than procedural accomplishments (such as the number of remedial investigations completed) that may even get in the way of broader cleanup goals; and elimination of indicators that conflict with, rather than contribute to, effective management.

Particular care should be taken to ensure that performance measures do not inadvertently encourage selection of one Superfund implementation strategy over another. If EPA regions are to be encouraged to make discerning strategic choices, the incentives system should be designed accordingly. This has thus far not been the case. For example, a referral to the Justice Department—the centerpiece action in the prosecution strategy—constitutes a countable "bean" in the EPA's system of regional performance measures, while no analogous milestone (short of the final consent decree that usually ends both prosecution and accommodation cases) exists for rewarding pursuit of an accommodation strategy.

While modifications of the performance measurement system within the EPA may encourage behavior more conducive to site remediation, there is a broader point to be made. We suspect that one of the most profound economic implications of the current system of incentives within the EPA is its impact on the aggregate societal costs incurred in Superfund cleanups—in terms of both actual cleanup expenses and transaction costs. Governmental actors bear few, if any, negative consequences of increasing either the remedial or the transaction costs borne by PRPs. Indeed, as discussed above, the most serious negative consequences facing Superfund personnel are those that would follow an inappropriate concession to PRPs, the proverbial "sweetheart deal." This situation is what encouraged the EPA to pursue its strategy of suing only a few "slam-dunk" PRPs in the Laskin Poplar and Tybouts Corner cases—leaving it to the private parties to obtain information on other participants through third-party litigation. It is what allows the agency to profess no interest whatever in the overall cleanup costs incurred by PRPs or how those costs are apportioned among the various parties—a situation richly illustrated by the fact that the EPA does not even maintain a system for collecting information on the costs incurred by private parties at Superfund sites.[20]

Other governmental actors, such as Justice Department attorneys and state environmental officials, can also impose costs on the cleanup process at little hardship to themselves—a point discussed previously. This leads to a variant of what economists call market failure, a situation in which the rational, self-interested behavior of individuals has deleterious societal consequences.

Some of the structural and organizational modifications suggested above could reduce the capacity of non-EPA actors to add gratuitously to cleanup and transaction costs. A reorganization of the internal system of incentives and bean counts within the EPA might also rationalize some aspects of bureaucratic behavior within the agency. But a broader problem remains: the basic Superfund scheme, based as it is on a liability standard that allows virtually unlimited costs to be transferred to PRPs, provides no compelling incentive for *any* government official to seek to minimize the aggregate societal resources committed to hazardous waste cleanups—at least at those sites where viable PRPs can be expected to foot the cleanup bill.

The government ultimately absorbs some costs, and government negotiators typically concede on some issues of concern to the PRPs. This is one of the major lessons of our case studies. But these concessions are made only to secure an agreement. In none of the cases we studied were concessions made in the basic principle that PRPs could (and perhaps should) ultimately be held liable for all cleanup costs determined by the agency to be necessary. This stance influences all agency relations with PRPs, regardless of the strategy adopted by the agency in implementing Superfund. It almost certainly serves to engender PRP perceptions of unfair, arbitrary decisionmaking on the part of government officials.

It is not clear what could be done to address this problem—if it is a problem—short of altering the Superfund liability scheme. One modest improvement would be for the EPA to establish an information system to keep track of all expenses at Superfund sites, including cleanup and transaction costs incurred by government units *and* PRPs. At a minimum, such data would inform the growing national debate on the societal costs of the Superfund program. More optimistically, confrontation with clear evidence of the cost implications of their decisions might lead government officials to be somewhat more circumspect.

However, so long as the unambiguous standard of strict, joint and several liability makes any viable PRP liable for all costs associated with site cleanup, no matter how minimally involved, EPA and Justice Department officials will have a clear and direct incentive—if not, as in the prosecution approach, a moral imperative—to impose theoretically unlimited costs on such parties, regardless of the dictates of equity or proportionality. And, so long as all cleanup and transaction costs are expected be borne by others (even if, as we have shown, this seldom actually occurs), government officials will have no direct incentive to try to contain or reduce these costs. We therefore move to a discussion of Superfund's liability scheme.

Superfund's Liability System

The liability system in Superfund profoundly influences almost every aspect of the program. Without question it is the program's most controversial element. Assigning liability lengthens the decision-making process and generates substantial transaction costs.[21] The dictates of strict, joint and several liability serve as a major criteria against which the EPA and others measure success in negotiations. The liability doctrine affects overall program organization, the EPA's conception of the problems it faces at specific sites, and the procedures the agency uses in dealing with PRPs. Differing perspectives on how to approach the liability of PRPs, we have noted, is a major factor distinguishing the prosecution, accommodation, and public works strategies for implementing the Superfund program.

Civil Liability as a Policy Tool

Our case studies suggest that there are several broader implications of the use of a liability scheme such as that found in the Superfund program. First, when the major legislative endowment to an agency comes in the form of legal rights, government officials are encouraged to take a "conservator" view of their store of those rights. Most bargaining in other contexts among experienced parties, or "repeat players,"[22] proceeds with an eye toward how resolution of a current dispute will affect future negotiations.[23] But the prospect of setting binding legal precedents elevates these concerns. Thus we observed, particularly in our prosecution cases, an effort by government lawyers to hold legal ground, to scrutinize all decisions in depth (which often means at length) for their implications for precedent, and to mistrust departures from past policies.[24]

Second, reallocations of legal rights and duties encourage a fault-based view of social problems. Most liability systems applied to a public policy problem involve establishing that someone is at fault for the existence of the problem; accompanying that determination is the requirement that the responsible party pay to compensate or correct for any resulting harms. The notion in environmental law that polluters should pay to correct the results of their prior activities is analogous in this sense to the products-liability doctrine requiring manufacturers of faulty products to pay compensation for injuries their products cause. Both of these uses of civil law focus more on distributing the costs of injuries in a socially beneficial way than on traditional ideas of fault or responsibility. But notions of fault and blame constitute a kind of moralistic baggage that inevitably accompanies uses of tort law, both from the perspective of involved parties and that of critics.[25]

Finally, the reallocation of costs from public to private sources shifts the focus of government decisionmaking away from a calculation of the worth of a given policy objective, measured by the public's willingness to pay for it, to an assessment of the costs of securing the compliance of those who are required to foot the bill.[26] Many policy tools, such as taxing and spending or incentives, establish the worth of an objective by balancing its cost against public willingness to pay for it. But in liability systems, as with regulation, monetary payment comes from those who fall into the legally defined net, while decisions about what to require of them are made by government. The locus of these decisions shifts from how much of a given good we can afford, to how much it will cost government in both economic and political terms to impose these costs on private parties.[27]

We suspect that a conservator view of government rights, a fault-based conception of policy problems, and a tendency to distort cost-benefit calculations are all potential impediments to rational policymaking when civil liability doctrine is used as a policy tool. These pathologies are magnified in the context of Superfund's most potent legal weapon, the doctrine of joint and several liability.

Joint and Several Liability in Superfund

Joint and several liability is the cornerstone of the Superfund liability scheme.[28] JSL was not explicitly part of either the original statute authorizing Superfund (the Comprehensive Environmental Response, Compensation, and Liability Act of 1980), nor was it mentioned specifically in either of the two subsequent reauthorization acts. Rather, JSL became part of the Superfund program as a result of a series of judicial decisions, most notably *U.S.* v. *Chem-Dyne Corporation*.[29]

A doctrine that theoretically holds the depositor of a teacup of mildly toxic substances into a massive landfill liable for all expenses incident to its remediation—regardless of fault or negligence, and independent of any notion of proportionality or a fair-share allocation of costs—requires some justification. The potential for injustice is substantial, legislative intent is unclear at best, and the substantive reasons for its application are not intuitively obvious.

Proponents have put forward three rationales for application of joint and several liability to PRPs in Superfund cases: JSL allegedly promotes informal settlements (the *settlement rationale*) and encourages PRPs to negotiate a satisfactory cleanup and avoid costly and time-consuming litigation; JSL gives the EPA leverage to force businesses with substantial resources to pay for cleanups, rather than leave the costs to the beleaguered taxpayer (the *deep-pocket rationale*); and JSL is said to have a positive influence on the future behavior of waste-generators, transporters, and disposers (the *incentive effects rationale*).

Much of the discussion of the validity of these three rationales has proceeded in the absence of any data; rather, in a manner common to law journals and economics reviews, the existence of these various behavioral effects has been more or less assumed.[30] We are not in a position to assess the incentive effects rationale, as this research did not examine waste-disposal practices.[31] Indeed, our research is not dispositive of any of these assertions. But we believe that it does lead to some tentative conclusions about the validity of the settlement and deep-pocket justifications for joint and several liability in Superfund.

The settlement rationale is based on the assertion that JSL serves as a powerful means of getting recalcitrant or potentially recalcitrant PRPs to the bargaining table. "Hiding in the weeds," in this view, becomes a perilous activity when the potential exists for an EPA lawsuit that could result in an enormous judgment for payment of cleanup costs. JSL might also be expected to reduce the incidence of legal bickering over PRP shares of cleanup expenses. In the words of former EPA Administrator Lee Thomas:

Knowledge that they [PRPs] can be held jointly and severally liable for full cleanup gives responsible parties the impetus to negotiate settlements for cleanups when the harm at a site is indivisible. Without this powerful tool, incentives for delay while parties quibble over the particulars of individual contribution at the site may outweigh the real priority—getting on with the job of cleanup.[32]

In light of the case studies constructed in this research, Thomas's statement cannot be read without some incredulity. It is undoubtedly true that the "club" of joint and several liability motivated the PRPs originally sued in Tybouts Corner and Laskin Poplar to stand up and take notice. But it is unclear how JSL reduced the "quibbles over the particulars of individual contribution" that resulted in years of costly legal maneuvering in these cases. Indeed, it is difficult to conceive of a process further removed from the "real priority" of "getting on with the job of cleanup" than the bickering and disputation among PRPs, and between PRPs and the government, that characterized these two cases. JSL may truly be a powerful tool; it may cause some PRPs, particularly those with deep pockets, to negotiate in good faith (or at least in better faith than would be the case in its absence). But JSL guarantees neither total nor timely success to the government in Superfund cases.

Our case studies also demonstrate several negative aspects to the doctrine, influencing the actions of both the PRPs and the government. We noted previously the potential for a rigid application of JSL to provoke resistance and obstruction on the part of PRPs who believe they are being treated unfairly. This reaction can be manifested in the legal arena, with a deluge of motions, third-party lawsuits, abusive discovery, and other delaying tactics. It can appear in the political arena, with end runs around the EPA to the courts or to friends on Capitol Hill and in the executive branch. It is thus not always clear that PRPs, faced with what they believe to be an outrageously unfair assessment or apportionment of cleanup costs, will behave as the rules of economic rationality might predict. Rather, the Superfund program is replete with evidence of behavior that illustrates the contrary reaction.

The impact of joint and several liability on the incentives and actions of government officials is seldom discussed. As we have already asserted, the theoretical potency of JSL leads government negotiators to take immoderate positions that ignite PRP resistance. Some in the Superfund program consider JSL to be a kind of moral standard; these individuals hold that any deviation from the rule of making PRPs pay for all costs, no matter how unreasonable this allocation may seem in commonsense terms, represents a sellout to business. Even those who do not hold this extreme position use the goal of a 100-percent PRP absorption of costs to gauge the success of negotiated settlements.

Our point here is simply that while JSL may provide settlement incentives to PRPs, it can also drive otherwise cooperative PRPs away from the negotiating table. And it creates incentives for government officials to engage in precisely the kind of quasi-prosecutorial behavior that is likely to produce such a result. When Congress reauthorized the Superfund program in 1986, it explicitly provided a number of settlement "carrots" designed to encourage PRPs to enter into good-faith negotiations to bring about speedy agreement without excessive transaction costs. As we have seen, however, joint and several liability, and the accompanying evaluative standard that holds concessions to PRPs to be unnecessary if not immoral, has led to underutilization of these provisions, and (if our case studies are any guide) additional delays and added transaction costs.

The deep-pocket rationale is subject to many of the same qualifications. While it is true that JSL suggests that the government can make disproportionately high cleanup costs the responsibility of wealthy corporations unlucky enough to find themselves among the PRPs at a Superfund site, our studies include cases in which deep-pocket PRPs remained outside of comprehensive settlements—as in Harvey and Knott Drum and Tybouts Corner. Furthermore, we observed no instances in which the depth of a PRP's pocket was the crucial factor in its ultimately negotiated share of cleanup costs. Some rough notion of proportionality animated the final apportionment in all the cases we studied. Our cases also illustrate that, regardless of the depth of PRPs' pockets and regardless of a liability doctrine that argues against its necessity, the government inevitably seems to ante up some part of the administrative (and sometimes, the cleanup) expenses and to absorb a portion of the future risks inherent in Superfund cleanups.

These ruminations suggest that the present liability standard in the Superfund scheme is, at the least, in need of reexamination. At the large number of sites that implicate only one PRP, joint and several liability is unimportant. JSL is most relevant at large sites with multiple PRPs, when a group of wealthy, vulnerable corporations can be proceeded against with little initial government effort, and when the

EPA expects that this group can then take on the responsibility for conducting the cleanup.[33]

Any modification of joint and several liability will bring with it a period of legal uncertainty, with accompanying delays while liability standards are litigated and clarified. The potential for yet more delay and uncertainty in the Superfund program argues for caution in altering ground rules that have been established at so great a cost, and over so lengthy a period of litigation. However, we do not regard JSL as an unmixed blessing, nor are its benefits as clear as its proponents sometimes assert. Ironically, as we suggest above, there are aspects of JSL that complicate the settlement process in Superfund cases, and encourage intransigence on the part of both PRPs and the government. Almost by definition, JSL also has considerable potential for unfairness and abuse. We are thus left with a suspicion that Superfund might actually work more effectively with a decreased reliance on joint and several liability.

One potentially fruitful change would be a stated legislative preference for a "fair-share" apportionment standard at multiple-party sites, with a clearer authorization for the government to cover the costs of orphan shares.[34] Because JSL was never explicitly incorporated in the statutory language, this change would not require substantial redrafting. A fair-share standard is, in fact, the informal criterion that reportedly underlies all Superfund settlements; it certainly was the basis of agreement in the cases we examined.

We suspect that use of a fair-share standard, particularly if it were combined with explicit legislative approval for allowing *de maximus* PRPs to buy out of future risks and liability at a site upon payment of a premium over the fair-share allocation, would not substantially increase the government's actual share of cleanup expenses. After all, the government would retain the potent coercive tools of unilateral administrative orders and the threat of treble damages. A fair-share standard would reduce substantially the potential for unfairness embodied in the version of JSL currently applied in Superfund cases. At least as important, it would alter the calculus and incentives of governmental actors, who would acquire a clear motivation to use the negotiating tools and settlement devices in the Superfund legislation, and to concern themselves with costs as well as the other values to be pursued at Superfund sites. Whether or not movement away from JSL would increase the incidence of PRPs' "hiding in the weeds" and avoiding the negotiating table cannot be determined without some experience with alternative liability schemes.[35]

Conclusion:
Homilies from Superfund on "Making Privatization Work"

At the risk of attempting to find the world in a grain of sand, we will in this final section draw some additional general conclusions relating the Superfund program to the ways that the United States has chosen to achieve an increasing number of its public policy goals.

Superfund and Privatization

Before proceeding with a discussion of the broadest, most conjectural, lessons of Superfund, we note that the unique elements of the Superfund program make some forms of generalization unproductive. Superfund's application of retroactive, strict, joint and several liability to environmental problems is, so far as we can ascertain, unique among the advanced industrialized nations of the world.[36] Even in the United States, these harsh liability doctrines are applied hesitantly, particularly in the area of public law. Other aspects of the Superfund program, however, are more typical of government programs in other policy domains. These include:

—Reliance on a regulatory apparatus to compel nongovernmental actors to deliver collective goods;

—Extensive use of private parties to achieve public policy ends;

—Creation of a process that requires an extensive system of sign-offs and requirements as a means of setting and achieving multiple goals.

The efforts of policymakers to deliver collective goods by regulating private behavior are well known. Fire safety codes for public buildings and health codes for restaurants are obvious examples. Because the credit for delivering a good is taken by the government, and the costs are displaced onto other parties, this approach is increasingly attractive in times of budgetary distress, in areas as disparate as national health insurance and the regulation of medical waste.[37] Why, it is asked, should we create a government agency to deliver a public good when the private sector can either be paid or coerced into providing it at (assumedly) lower cost and higher levels of performance?

Despite some wrinkles (such as the retroactive liability scheme), Superfund's use of government power to assign the costs of hazardous waste cleanup to private parties falls squarely within this general approach.

Similarly, Superfund exemplifies the increasing tendency of government to use private parties to deliver public goods and services ranging from vocational education to housing for homeless people. Privatization—paying people to do things government wants done— is found in Superfund's heavy reliance on contractors to conduct site investigations, engineering studies, cleanups, and monitoring activities. Indeed, the entire Superfund liability apparatus, divorced as it is from traditional notions of responsibility and fault, frequently is viewed as simply a means of forcing private parties to deliver the public good of environmental cleanup without compensation.

As more public activities are placed in private hands, particularly the hands of unwilling parties, government should be increasingly concerned over the quality of the goods and services delivered. To ensure that important social values are not neglected in this process, or to augment the public goods and services delivered, lawmakers create formal sign-off requirements to be completed by those officials who have become the formal guardians of particular public concerns. Hence the necessity of obtaining environmental impact statements for all major construction projects; meeting federal, state, and local ARARs when hazardous waste facilities are cleaned up; and complying with handicapped-access requirements on building permit approvals.

Taken together, these steps toward privatization have converted many government agencies from deliverers of public goods into enforcers of the law, managers of contracts, guardians of standards, and negotiators of deals. These agencies have been denied the means of achieving their goals through direct action, often because of fiscal or statutory constraints based on an orthodoxy of distrust of government-run programs. They act instead through nongovernment agents. Yet we continue to hold the government responsible for the achievement of an ever-growing set of designated outcomes. While the tools and actions of government have changed, sometimes profoundly, public expectations have not been modified accordingly. This disjuncture between expectations and reality, as we will argue, may have increasingly undesirable consequences.

Legislative policymakers have less control over outcomes in a privatized system because they must work indirectly. Because legislators seldom can command private parties to deliver public goods, they must rely on government agencies, which are expected to use the carrots and sticks provided them in statutes to obtain the desired private action. As has been made evident by the Superfund program, this indirect attempt to influence private behavior does not always produce the desired results. Government agencies such as the EPA, for their part, are increasingly held to performance standards that are not in their direct power to meet. And the fragmentation of the sign-off process increases transaction costs substantially.

Private parties, whether the beneficiaries of contracts or the targets of liability systems, find themselves at the center of frequently conflicting requirements originating from a variety of government agencies. Their responses vary from the pursuit of private self-interest at public expense to assumption of the full responsibility for resolving policy quandaries and contradictions not faced in the pubic sector.

Prescriptions for Successful Policy Privatization

Despite the thrust of our discussion thus far, we do not see ourselves as engaging in a jeremiad against the ways public goods are delivered in present-day America. Rather, assuming that these choices have been made for good reasons, and that this approach for better or worse is likely to continue, we ask how these arrangements might be made to work better. How can we push more policy outcomes produced by government reliance on private intermediaries toward the successful end of the continuum, where goods are delivered and programs succeed, and away from deadlock, exploitation, and policy failure? Based on a broad reading of the successes and failures of Superfund, we offer a few prescriptions that we regard as simple and reasonable, though perhaps also unpalatable.

—Consider transaction costs when selecting among the goals to be met, and take into account the costs paid by private actors as well as government. Policymakers must recognize that working toward multiple goals through private-party action will require substantial transaction costs. Government agencies therefore need to be staffed adequately to ensure that the requisite attention is paid. Currently, goals and sign-offs are added as if they were without costs (as in Superfund's ARARs requirements), private costs are not an official concern (a major theme of our Superfund portraits), and government staffing remains constant, based upon an implicit expectation that, because services are not being delivered by the government, relatively few public employees are needed to run the programs. While this perspective may appeal to budget officers in the executive branch and elsewhere, it may not lead to effective program implementation.

—*Recognize that the agency's capacity to fine-tune outcomes is limited when it does not perform the work.* The rule of thumb probably should be that the more removed an agency is from delivery of a public good, the less control policymakers can expect to exert. We currently accept this principle in some cases, but not in others. Thus, federal student aid has not been used to calibrate college curricula and graduation standards, but we do expect federal power to determine both the specific means and precise outcome standards for privately conducted Superfund cleanups. Public policy problems are frequently regarded as though the principal-agent problem does not exist.

—*Recognize the times when "the best is the enemy of the good."* Too frequently, government policies focus on optimal solutions, to the detriment of a realistic assessment of what might generate the greatest good for the greatest number. This problem is particularly acute in a privatized intermediary system, in which private incentives and decisions heavily influence the final results of government initiatives. For example, our approach to automobile pollution control has been to apply stricter standards to new cars while largely ignoring old ones; our approach to improving the lot of people with disabilities has been in large part to apply stringent access requirements to new construction and renovation and leave old structures alone.

At times, placing heavy requirements on what is new pushes people into more use of what is old. Thus, in Superfund, making the settlement process more cumbersome by raising standards and diminishing private incentives to cooperate will likely increase PRP recalcitrance and the incidence of hiding in the weeds. Rather than seek the "best" settlements, we might want to put more emphasis on achieving "good" ones. Calibrating these generalizations is, of course, no small task. But it is not plainly more difficult than making the present system work.

—*An image of reasonableness in the enforcement of the government's legal prerogatives may be more productive of compliant behavior than an emphasis on coercion.* We are not the first to suggest that, despite the presence of potent coercive tools, government actions perceived to be unreasonable can cause resistance, rather than the begrudging compliance that might be predicted by a simple cost-benefit calculus.[38] It may be that our public culture embraces the fatherly advice to "pay the five dollars," despite the alleged unfairness of arbitrary action by a public official. But this is also the land where many would "rather be right than president." The Superfund program amply demonstrates the naiveté of assuming that if government is tough and coercive enough, and holds all the legal cards, it can produce speedy acquiescence by the private parties whose acts or forbearance will further the public good. The experience of regulatory programs in other policy areas, such as occupational health and safety regulation, points to a similar conclusion.

As a related observation, we also suspect that the effect of joint and several liability on the incentives and behavior of all those government officials associated with the Superfund program may find corollaries elsewhere. As stated in the previous section, we have a nagging suspicion that the very potency of this legal club may have encouraged Superfund lawyers and administrators to engage in behavior contrary to the overall goal of attaining speedy, voluntary site remediation. JSL became a kind of defining standard for what should be obtained in Superfund settlements within the EPA, in Congress, and in much of the environmental community. This standard runs contrary to the aspects of the program that focus on conciliation, cooperation, and voluntary agreements with PRPs. Put in a different policy context, the lesson may be to avoid giving law-enforcement officers flack jackets, attack dogs, and shoot-to-kill orders if what you really want is better police-community relations and more voluntary compliance by citizens with their legal obligations.

We have taken the prerogative of scholars—individuals with some knowledge but little power—to offer these last observations. They are hardly the stuff of new paradigms or the nuts-and-bolts of specific programmatic assistance. But, like Yogi Berra, we are less concerned that society will make mistakes than we are with the prospect of living with the consequences of having made the wrong mistakes.

Glossary

ARAR Applicable or relevant and appropriate regulation. An environmental, health, or other standard that must be met in a Superfund cleanup. These standards typically concern air, water, or soil cleanliness, and may be set by localities, states, branches of the EPA, or other components of the federal government (such as the Fish and Wildlife Service or the Coast Guard).

CERCLA Comprehensive Environmental Response, Compensation, and Liability Act of 1980 (P.L. 96–510). The statute establishing the Superfund program.

contribution A legal doctrine that enables parties sued under joint and several liability to obtain compensation from other parties who may have been legally liable, but who were not proceeded against in the original court action.

cost recovery action A legal proceeding, authorized under CERCLA, that allows the government to proceed against PRPs for recovery of both administrative and actual cleanup costs expended in either emergency removal or remedial activities at hazardous waste sites.

delisting The process by which a Superfund site is removed from the National Priorities List (NPL) after it has been completely cleaned up.

de maximis **parties** Parties at a Superfund site that have primary legal responsibility for its cleanup and adequate resources to fulfill that responsibility.

de minimis **buyout** A practice, authorized in SARA, which allows the government to accept a fixed monetary sum from a PRP in exchange for a complete end to liability at a Superfund site. This provision is available only to smaller contributors of waste to a site.

EPA United States Environmental Protection Agency. The federal agency charged with enforcement of the Superfund program. Throughout this book, the acronym *EPA* refers to the federal Environmental Protection Agency. State agencies (such as the Ohio EPA) are prefaced by the state name.

JSL Joint and several liability. The legal doctrine that specifies that any one defendant in certain tort cases can be held liable for *all* the damages for injuries caused by the actions of a group of several defendants.

mixed funding The practice, authorized in SARA, by which the government can assume some proportion of cleanup expenses, with other parties assuming the rest. According to current policy, it is used only when the government's share can be obtained in subsequent cost recovery actions against nonsettling parties.

NBAR   Nonbinding assessment of responsibility. A device, established in SARA, that allows the EPA to make a nonbinding estimate of the proportional share that each of the various responsible parties at a Superfund site should pay toward the costs of cleanup.

NPL   National Priorities List. The list of approximately twelve hundred hazardous waste sites that have been determined (through a hazardous ranking score) to constitute the most serious environmental threats in the United States.

O&M costs   Operating and maintenance costs. The expenses of maintaining an ongoing engineering remedy at a Superfund site. Depending on the remedy chosen, these costs can range from very low to extremely expensive. They can continue to be generated for decades.

operable unit   A division of a larger Superfund site. The practice of establishing operable units frequently allows action to proceed on parts of a cleanup incrementally, rather than be delayed until all issues are resolved and all aspects of a remediation can go forward at the same time.

ORC   Office of Regional Counsel. The EPA's legal office in the regions. Typically, an ORC attorney is assigned to each Superfund case.

orphan share   The term used to describe the share of waste at a site that cannot be collected from a PRP, because the PRP is either unidentifiable or insolvent.

petroleum exclusion clause   A clause in CERCLA that exempts some petroleum wastes and their generators from the Superfund cleanup scheme.

PRP   Potentially responsible party. Either an individual, a business, or a government unit that has been identified as a party that is potentially liable for site cleanup under the provisions of CERCLA.

RCRA   Resource Conservation and Recovery Act of 1976 (P.L. 94–580). The act, administered by the EPA, that regulates the definition, transportation, and disposal of hazardous wastes. This act is distinct from the Superfund statutes in that it regulates current and future waste disposal practices, while Superfund was established to cleanup inactive hazardous waste dumps.

RD/RA   Remedial design/remedial action. The final stage of a site cleanup, when the remedy is designed and put into effect.

remedial procedures   The complex, statutorily mandated procedures for decisionmaking at nonemergency Superfund sites. These include the RI/FS, ROD, and RD/RA.

remediation   Site cleanup.

removal   Or, *emergency removal*. An action taken by the EPA under the emergency removal provisions of CERCLA, that enables the agency to take preliminary steps to clean up a site or reduce its danger when there is an imminent and substantial threat to public health or the environment. The cost of a removal cannot exceed $2 million for any one action at any one site. See also *removal*.

Removial The controversial practice of using the emergency removal authority of the EPA (see *removal*) at sites where remedial action (see *remedial procedures*) is more appropriate. Removials may involve exaggeration of the threat posed by a site, or conscious underestimation of the cost of an action to fit it under the $2-million cap on removals. Removials can also involve attempts to break a larger site into smaller operable units, each satisfying the $2-million limit.

reopener A clause, usually included in Superfund consent decrees at government insistence, which allows the government to reopen a case and proceed legally against a responsible party who has already settled with the government if certain contingencies of site cleanup (such as discovery of additional, unexpected waste, or failure of a remedy) occur.

RI/FS Remedial investigation/feasibility study. The *remedial investigation* is an engineering study that assesses the geographical, geological, and hydrological properties of a site, and the nature and extent of the hazardous waste contained therein. It is usually combined with the *feasibility study*, which identifies the various cleanup alternatives and specifies their costs and benefits.

ROD Record of decision. The formal document by which an EPA administrator (usually the regional administrator) chooses the remedy to be applied at a Superfund site. The ROD is listed in the *Federal Register* within the requisite comment period.

RPM Remedial project manager. The EPA official, often a recent graduate in engineering or science, who has charge of the remediation at a particular Superfund site.

SARA Superfund Amendments and Reauthorization Act of 1986 (P.L. 99–499). The act reauthorizing the Superfund program and adding a number of additional provisions, such as several incentives to encourage voluntary settlements, as well as the requirement that Superfund cleanups meet applicable or relevant and appropriate regulations (see *ARAR*).

section 106 order A unilateral administrative order, authorized by statute, that allows the EPA to order PRPs to perform certain remedial actions at a Superfund site, subject to treble damages and daily fines if the order is not obeyed.

strict liability The legal doctrine that allows a defendant in certain tort cases to be held liable for injuries, regardless of whether or not that party was negligent.

Superfund (1) The revolving trust fund, maintained both from general federal revenues and a tax on chemical manufacturers, that the EPA draws on to pay for site cleanups when no viable responsible parties can be found, or when viable responsible parties refuse to cooperate. The fund can be replenished by private-party contributions obtained through settlement negotiations or lawsuits. (2) A shorthand expression for the entire hazardous waste cleanup program established by CERCLA.

third-party suits   In the context of Superfund, third-party suits are those brought by PRPs at a site who are sued by the government, and against other PRPs who were not sued, in order to obtain compensation for their costs and expenses. See *contribution*.

viable PRP   A PRP who can be expected to pay his or her share of cleanup costs.

waste-in list   A list, sometimes prepared by regional EPA officials, that identifies the amount and nature of the waste deposited at a site by each of the PRPs.

Notes

Notes to Preface

1. The company is the American Insurance Group. See "America needs a new system to achieve fast and effective cleanup of our environment," advertisement in *Newsweek*, October 14, 1991, pp. 74–75. AIG and other insurers have produced a series of case studies critical of the liability doctrine and the transaction costs associated with cleanups, see, for example, American Insurance Group, "Superfund Issues Forum: Superfund Site Study of Picillo Farm, Coventry, Rhode Island," Washington, 1991.

2. See, for example, Daniel Mazmanian and David Morell, *Beyond Superfailure: America's Toxics Policy for the 1990s* (Westview Press, 1992); M. R. English and others, *The Superfund Process: Site-Level Experience* (University of Tennessee, Waste Management Research and Education Institute, 1991); Environmental Law Institute, "ELI Draft Enforcement Report," Washington, 1989; Frank Lautenberg and Dave Durenberger, *Lautenberg-Durenberger Report on Superfund Implementation: Cleaning Up the Nation's Cleanup Program* (Washington: Senate Subcommittee on Superfund, Ocean, and Water Protection, 1989); Office of Technology Assessment, *Are We Cleaning Up? Ten Superfund Casestudies* (Washington, 1988); Office of Technology Assessment, *Coming Clean: Superfund's Problems Can Be Solved* (Washington, 1988); Clean Sites, *Improving Remedy Selection: An Explicit and Interactive Process for the Superfund Program* (Alexandria, Va., 1990); Environmental Defense Fund and others, *Right Train, Wrong Track: Failed Leadership in the Superfund Cleanup Program* (Washington, 1988); Jan Paul Acton, *Understanding Superfund: A Progress Report* (Santa Monica, Calif.: Rand Institute for Civil Justice, 1989); Center for Hazardous Waste Management, Illinois Institute of Technology, *Coalition on Superfund Research Report* (Washington: Coalition for Superfund, 1989).

3. Peter Passell, "Experts Question Staggering Costs of Toxic Cleanups," *New York Times*, September 1, 1991, p. A1.

4. Of the 1,236 sites in 1989, cleanup was under way at 272 and completed at another 63, see "Breakdown of Superfund Activity," *National Law Journal*, February 18, 1991, p. 39.

5. Selection of the appropriate technology to effect the cleanup is certainly an important element in this process, but we will not focus on the selection of remedy in any detail, nor will we attempt to assess whether the remedy chosen is legally sufficient, scientifically defensible, or otherwise more or less

desirable than alternatives. Several studies have recently been completed of this aspect of the Superfund program. See Clean Sites, *Improving Remedy Selection;* and Environmental Defense Fund, *Right Train, Wrong Track.*

6. Environmental Law Institute, "ELI Draft Enforcement Report"; and Lautenberg and Durenberger, *Lautenberg-Durenberger Report on Superfund Implementation.*

7. American Telephone and Telegraph and others, "Superfund from the Industry Perspective: Suggestions to Improve and Expedite the Superfund Remediation Process" (Washington, 1989).

8. Case studies of Superfund cleanups can be found in English and others, *The Superfund Process.* The Coalition for Superfund financed an earlier study conducted by Illinois Institute of Technology that was designed to examine enforcement issues in individual cases; this aspect of the study was not pursued because of the difficulty researchers faced in obtaining "enforcement confidential" information from government attorneys. See Center for Hazardous Waste Management, *Coalition on Superfund Research Report.*

Chapter One

1. The entire program, including the accompanying taxes, was extended by four years as part of the Omnibus Budget Reconciliation Act of 1990, P.L. 101–508, section 6301. For the figure on tax dollars see Peter Guerrero, "Superfund: Issues That Need to Be Addressed before the Program's Next Reauthorization," Testimony before the Subcommittee on Investigations and Oversight of the House Committee on Public Works and Transportation, October 29, 1991, 102 Cong. 1 sess. (Government Printing Office, 1991).

2. The authorizing statutes are CERCLA—the Comprehensive Environmental Recovery, Compensation, and Liability Act of 1980, P.L. 96–510 —and SARA—the Superfund Amendments and Reauthorization Act of 1986, P.L. 99–499. For a discussion of the passage of Superfund, see Marc K. Landy, Marc J. Roberts, and Stephen R. Thomas, *The Environmental Protection Agency: Asking the Wrong Questions* (New York: Oxford University Press, 1990), and Bong H. Lee, "Shifting Gears in a Dynamic Environment: Changing Strategies in the Formulation of Superfund," Ph.D. dissertation, State University of New York at Albany, 1991.

3. See Clean Sites, *Making Superfund Work: Recommendations to Improve Program Implementation* (Alexandria, Va., 1989).

4. This figure includes other, smaller, environmental cleanup programs, such as those dealing with underground tanks and nuclear waste. These researchers' "best guess" of aggregate cleanup costs of all of these programs under current EPA policy was $750 billion. Milton Russell, E. William Colglazier, and Mary English, *Hazardous Waste Remediation: The Task Ahead* (University of Tennessee, Waste Management Research and Education Institute, 1991), Executive Summary.

5. The 1990 reauthorization of the Superfund statutory scheme was ac-

complished under the Omnibus Budget Reconciliation Act of 1990. This legislation merely extended the expiration dates of the program and accompanying taxes; no substantive changes were made in the program.

6. This process is described at length in Jan Acton, *Understanding Superfund: A Progress Report* (Santa Monica, Calif.: Rand Institute for Civil Justice, 1989) pp. 11–17.

7. However, these requirements sometimes have been honored in the breach. By bending of the dangerousness criterion, and by creative accounting practices, sites arguably inappropriate for removals have managed to meet removal requirements, and thus avoid the more cumbersome and time-consuming procedures necessary in nonemergency remedial cleanups. See the discussion of Region IV's use of "removials" in chapter 5.

8. Emergency conditions at the site can be handled by means of removal actions, and many of the largest cleanups began with emergency actions to neutralize a site's most dangerous aspects.

9. As is shown in part 2, the practices of the various EPA regions differ in relation to their interest in bringing PRPs "on board" for the RI/FS process.

10. This assessment may be based on everything from hunches to explicit agreements with PRPs concerning the preferred remedial alternative. Compare, for example, the role of PRPs in remedy selection in the Tybouts Corner case in Region III to that role in Laskin Poplar in Region V; the cases are discussed in chapters 4 and 3, respectively.

11. For a brief but cogent discussion of the major scientific and engineering techniques and concepts used in Superfund cleanups, see Melvyn Kopstein, "Science for Superfund Lawyers," *Environmental Law Reporter*, vol. 19 (September 1989), pp. 10388–92.

12. These figures are based on data collected during the first eight years of the program, so the total elapsed time cannot exceed eight years. This fact suggests that average time may increase as the program continues. However, early Superfund cases may have consumed more time in what was a "shakedown" period for the EPA. Acton, *Understanding Superfund*, p. 16.

13. The EPA estimates that the twelve hundred sites currently on the NPL will cost about $40 billion to clean up. This works out to a little over $33 million per site. See U.S. General Accounting Office, *Superfund: More Settlement Authority and EPA Controls Could Increase Cost Recovery* (Washington, July 1991), p. 9.

14. For a historical review stressing "enforcement" as a central approach, with other approaches as deviations, see Joel A. Mintz, "Agencies, Congress, and Regulatory Enforcement: A Review of EPA's Hazardous Waste Enforcement Effort 1970–1987," *Environmental Law*, vol. 18 (Summer 1988), pp. 683–777. See also Harold Barnett, "Political Environments and Implementation Failures: The Case of Superfund Enforcement," paper presented at the 1989 annual meeting of the Law and Society Association, Madison, Wis.

15. See Eugene Bardach and Robert Kagan, *Going by the Book: The Problem of Regulatory Unreasonableness* (Temple University Press, 1982).

16. Environmental Law Institute, "ELI Draft Enforcement Report" (Washington, 1989), sec. 2, p. 1.

17. Ibid., p. 2; and Frank Lautenberg and Dave Durenberger, *Lautenberg-Durenberger Report on Superfund Implementation: Cleaning Up the Nation's Cleanup Program* (Washington: Senate Subcommittee on Superfund, Ocean, and Water Protection, 1989).

18. The term is from Steven Kelman, "Adversary and Cooperationist Institutions for Conflict Resolution in Public Policymaking," *Journal of Policy Analysis and Management*, vol. 11 (Spring 1992), pp. 178–206.

19. There is a growing literature describing the use of alternative dispute resolution in the area of environmental regulation. For an example of one unsuccessful effort to apply alternative dispute resolution to the cleanup of a toxic waste site, see Robert T. Nakamura, Thomas W. Church, and Phillip J. Cooper, "Environmental Dispute Resolution and Hazardous Waste Cleanups: A Cautionary Tale of Policy Implementation," *Journal of Policy Analysis and Management*, vol. 10 (Spring 1991), pp. 204–21.

20. For a historical review stressing enforcement as a central approach, and other approaches as deviations, see Joel A. Mintz, "Agencies, Congress, and Regulatory Enforcement: A Review of EPA's Hazardous Waste Enforcement Effort 1970–1987," *Environmental Law*, vol. 18 (Summer 1988), pp. 683–777.

21. The American Insurance Association and the American Insurance Group have proposed that the retroactive liability aspects of Superfund be eliminated, to be replaced by publicly conducted cleanups, financed by a tax on liability insurance.

22. Offices dominated by engineers reportedly took more of a public works orientation; lawyer-administrators were said to prefer prosecutorial or accommodation approaches.

23. Both CERCLA and SARA require the state to pay a share of the cost of cleanups financed by Superfund revenues. If a state or municipality is involved as a PRP (e.g., if a locality operates a landfill found to contain hazardous wastes), this share is much higher.

24. Region V leads the country, with 266 NPL sites. It is followed by Region II (201 sites), Region IV (155 sites), and Region III (152 sites). Source: EPA CERCLIS data base.

25. Officials in Regions III and V seemed to agree with the characterizations of their respective approaches as accommodation and prosecution. Region IV officials indicated that while a public works perspective may have been taken in their region in the past, their approach was now far more enforcement-oriented. We are interested, however, in the impact of different implementation strategies rather than the present situation in any EPA region. This changing regional perspective, however, complicated case selection in Region IV.

26. "I shall not today attempt to define the kinds of material I understand to be embraced within that shorthand description [obscenity], and per-

haps I could never succeed in intelligently doing so. But I know it when I see it." Justice Potter Stewart concurring in *Jacobellis* v. *Ohio*, 387 US 184 (1964).

27. Interview notes were transcribed immediately after each interview. These transcriptions served as the basis of all quotations reported in the text. Because we felt that a tape recorder would have inhibited some interviewees, our records of the interviews are based on handwritten notes. Transcriptions of these notes serve as the basis of the quotations reported throughout this study. When interview subjects are quoted, the words reported are as close as possible to the actual words used.

Chapter Two

1. For representative examples of this new genre, see Eugene Bardach, "Implementation Studies and the Study of Implements," paper presented at the 1980 annual meeting of the American Political Science Association; Martin A. Levin and Barbara Ferman, *The Political Hand: Policy Implementation and Youth Employment Programs* (Pergamon Press, 1985); Lorraine M. McDonnell and Richard Elmore, "Getting the Job Done: Alternative Policy Instruments," *Educational Evaluation and Policy Analysis*, vol. 9 (Summer 1987), pp. 133–52; Lester Salamon, ed., *Beyond Privatization: The Tools of Government Action* (Washington: Urban Institute Press, 1989); and Anne Schneider and Helen Ingram, "Behavioral Assumptions of Policy Tools," paper presented at the 1989 annual meeting of the Midwest Political Science Association.

2. McDonnell and Elmore, "Getting the Job Done," p. 133.

3. The operant legislation included the Atomic Energy Act of 1950 and the Anderson-Price Atomic Energy Damages Act of 1957. See National Childhood Vaccine Injury Act of 1986; and the Vaccine and Immunization Amendments of 1990.

4. Civil Rights Commission Authorization Act of 1979.

5. While advocates sometimes argue that Superfund can encourage businesses to deal with hazardous wastes more carefully, or even to engage in the "midnight cleanup" of inactive sites that are not yet listed, such impacts are speculative and no good evidence yet exists that they occur for that reason. For a discussion of indirect benefits, see Joint Testimony (of Clean Water Action, Environmental Defense Fund, Friends of the Earth, Greenpeace, Natural Resources Defense Counsel, Sierra Club, U.S. Public Interest Research Group) prepared by Douglas Wolf and delivered in Hearings before the Subcommittee on Investigations and Oversight on Issues Relevant to the Reauthorization of the Superfund Statute of the House Committee on Public Works and Transportation, 102 Cong. 1 sess. (Government Printing Office, 1991). After noting a number of possible benefits, the testimony calls for further research to document them. The difficulties of documenting such bene-

fits is discussed in Resources for the Future, "Summary of RFF-EPA Superfund Economics Forum" (memo to conference attendees), Washington, April 6, 1992, pp. 2–4.

6. Donald Clay, EPA's assistant administrator in charge of Superfund, reiterated the agency's commitment: "We are firmly committed to the law's 'polluter pays' approach to clean up. As a result, over 60% of all cleanups under way are being taken by responsible parties, up from 37% four years ago. These cleanups are now occurring through newly achieved record levels of lawsuits and enforcement orders. The total value of private-party commitments now exceeds $5 billion, including a record $1.4 billion this year alone." EPA Communications and Public Affairs, Note to Correspondents, November 26, 1991.

7. For a discussion of such constraints or "imperatives," see Martin Rein and Francine Rabinovitz, "Implementation: A Theoretical Perspective," in W. E. Burnham and M. W. Weinberg, eds., American Politics and Public Policy (MIT Press, 1978).

8. A notable exception to the dearth of literature on implementation strategies for the use of other policy tools is Charles Levine, Guy Peters, and Frank Thompson, Public Administration: Challenges, Choices, Consequences (Harper and Collins, 1991). See especially chap. 11, which deals with the public management challenges associated with different statutory "technologies."

9. For a discussion of the role of presentation in shaping the behavior of others, see Erving Goffman, The Presentation of Self in Everyday Life (Woodstock, N.Y.: New York Overlook Press, 1973).

10. Actually, this lawyer used the term to describe perhaps the most potent tool in the EPA's arsenal, the doctrine of joint and several liability. We have expanded its original meaning somewhat in this discussion. See William Wilkerson and Thomas Church, "The Gorilla in the Closet: Joint and Several Liability and the Cleanup of Toxic Waste Sites," Law and Policy, vol. 11 (October 1989), pp. 425–49.

11. Other researchers have found analogous regional differences in enforcement of the Clean Air Act. See Susan Hunter and Richard W. Waterman, "Determining An Agency's Regulatory Style: How Does the EPA Water Office Enforce the Law?" Western Political Quarterly, vol. 45 (June 1992), pp. 403–18.

12. For a general discussion of the mandating or regulatory policies, see Eugene Bardach and Robert A. Kagan, Going by the Book: The Problem of Regulatory Unreasonableness (Temple University Press, 1982). Some of the same issues are dealt with in a broader context in McDonnell and Elmore, "Getting the Job Done: Alternative Policy Instruments."

13. Congressional Record, October 3, 1986, p. S14903. Cited in Frank Lautenberg and Dave Durenberger, Lautenberg-Durenberger Report on Superfund Implementation: Cleaning Up the Nation's Cleanup Program (Washington: Senate Subcommittee on Superfund, Ocean, and Water Protection, 1989), p. 31.

14. Lautenberg and Durenberger, Lautenberg-Durenberger Report, p. 33.

15. Environmental Law Institute, "ELI Draft Enforcement Report" (Washington, 1989), sec. 2, pp. 16, 18.

16. Environmental Protection Agency, *Progress toward Implementing Superfund* (Washington, June 1991), chart titled: "Superfund Program Highlights: Myth versus Reality."

17. The policy-machine image is from Eugene Bardach, *The Implementation Game: What Happens after a Bill Becomes a Law* (MIT Press, 1977).

18. Robert Kagan and John Scholz, "The 'Criminology of the Corporation' and Regulatory Enforcement Strategies," in Keith Hawkins and John Thomas, eds., *Enforcing Regulation* (Kluwer/Nijoff, 1984), pp. 67–95.

19. The term is from Kagan and Scholz, "The 'Criminology of the Corporation' and Regulatory Enforcement Strategies," pp. 69–74.

20. American Telephone and Telegraph and others, "Superfund from the Industry Perspective: Suggestions to Improve and Expedite the Superfund Remediation Process" (Washington, February 1989), pp. 1–2. See also Kimberly Ann Leue, "Private Party Settlements in the Superfund Amendment and Reauthorization Act of 1986," *Stanford Environmental Law Journal*, vol. 8 (1989), pp. 131–73.

21. Indeed lawyers in Region III told us that they considered the program staff their clients.

22. See Kagan and Scholz, "The 'Criminology of the Corporation' and Regulatory Enforcement Strategies," pp. 74–80.

23. The cleanup programs of Denmark, the Netherlands, and Germany all have strong public works components, although each also attempts (to a much lesser degree than in the United States) to obtain private-party participation at sites where responsibility is clear. See Andrew Lohof, "The Cleanup of Inactive Hazardous Waste Sites in Selected Industrialized Countries," Discussion Paper 069 (Washington: American Petroleum Institute, 1991); Raymond Kopp, Paul Portney, and Diane DeWitt, "International Comparisons of Environmental Regulation," Discussion Paper QE90-22-REV (Washington: Resources for the Future, September 1990).

24. It is not accidental that Region IV makes such extensive use of removals, given that it has been one of the few regions with operational RCRA sites suitable for accepting wastes from Superfund sites.

25. John Braithwaite, *To Punish or Persuade: Enforcement of Coal Mine Safety* (State University of New York Press, 1985).

26. See Braithwaite, *To Punish or Persuade*; Neal Shover, Donald Clelland, and John Lynxwiler, *Enforcement or Negotiation: Constructing a Regulatory Bureaucracy* (State University of New York Press, 1986); Hunter and Waterman, "Determining an Agency's Regulatory Style"; Robert Kagan, "Adversarial Legalism and American Government," *Journal of Policy Analysis and Management*, vol. 10 (Summer 1991), pp. 369–406; Steven Kelman, "Adversary and Cooperationist Institutions for Conflict Resolution in Public Policymaking," *Journal of Policy Analysis and Management*, vol. 11 (Spring 1992), pp. 178–206; John M. Mendeloff, *Regulating Safety: An Economic and Political Analysis of Occupational Safety and Health Policy* (MIT Press, 1979).

27. Kagan, "Adversarial Legalism and American Government." A more

cooperationist view is advocated in Kelman, "Adversary and Cooperative Institutions for Conflict Resolution in Public Policymaking."

28. This line of argument is summarized in Bardach and Kagan, *Going by the Book*. See also Environmental Law Institute, "ELI Draft Enforcement Report."

29. Theodore Lowi, *The End of Liberalism* (Norton, 1989).

30. Grant McConnell, *Private Power and American Democracy* (Vintage, 1966).

31. The term *cooperationist* was coined by Kelman in "Adversary and Cooperationist Institutions for Conflict Resolution in Public Policymaking."

32. For thoughtful discussions of America as an adversarial society and the implications of this characterization, see Lawrence M. Friedman, *Total Justice* (New York: Russell Sage Foundation, 1985); Marc Galanter, "Reading the Landscape of Disputes: What We Know and Don't Know (And Think We Know) about Our Allegedly Contentious and Litigious Society," *UCLA Law Review*, vol. 31 (October 1983), pp. 4–71.

33. For example, Probst and Portney write: "Through the program's history, there has been a tension between using Fund money to get sites cleaned up quickly and using EPA's broad enforcement powers to encourage or compel private parties to clean up sites." Katherine N. Probst and Paul R. Portney, *Financing Superfund Cleanups: The Search for a Better Mechanism* (Washington: Resources for the Future, February 20, 1991), p. 1. Jan Acton makes a similar point when he says that the Superfund program's incentives "may run at cross purposes and significantly affect the cost and pace of its achievements." *Understanding Superfund: A Progress Report* (Santa Monica, Calif.: Rand Institute for Civil Justice, 1989), p. vi.

34. Superfund is a telling example of this after-the-fact process of justification. If no cleanup activity takes place, the EPA can claim that resources were husbanded; if trust moneys are exhausted, the EPA can claim much activity; if the program is bogged down in endless litigation, the EPA can claim this is necessary to minimize taxpayer expenses. Prosecutors operate in a similar environment of multiple goals and ambiguous performance measures. See Thomas W. Church and Milton Heumann, *Speedy Disposition: Monetary Incentives and Policy Reform in Criminal Courts* (State University of New York Press, 1992), chap. 5.

35. Our list differs from a recent one by Resources for the Future only insofar as that organization is more concerned with indirect effects and fairness. This difference can be attributed to differences between our respective tasks: we are assessing the performance of strategies in achieving goals at a specific site and Resources for the Future is looking at alternative liability systems. See Katherine N. Probst and Paul R. Portney, *Assigning Liability for Superfund Cleanups: An Analysis of Policy Options* (Washington: Resources for the Future, June 1992), p. ix. Probst and Portney use the following criteria: "speed of cleanup, transaction costs, voluntary cleanups at non-NPL sites, due care in future waste management practices, fairness, and financial implications."

36. The cost-effectiveness criterion has generated much debate. The latest iteration of the National Contingency Plan indicates that the EPA sees the cost-effectiveness standard as one ensuring that the remedy chosen in a record of decision provides "reasonable value for the money" (53 *Federal Register* 51429). Congressional critics see cost effectiveness only coming into play after a particular remedy has been chosen and the issue is which specific engineering technique to apply. See Lautenberg and Durenberger, *Lautenberg-Durenberger Report*, pp. 57–64. The Superfund statute also requires that the EPA gives preference to treatment which permanently and significantly reduces the volume, toxicity, or mobility of hazardous substances. Sec. 121, P.L. 96-510, as amended by P.L. 99-499.

37. Clean Sites, *Main Street Meets Superfund* (Alexandria, Va., 1992).

38. 42 U.S.C. Sec. 9616 (1988).

39. This definition of transaction costs is somewhat narrower than the standard economics definition as "the costs incurred in negotiating and completing a transaction." Richard Lipsey, Peter Steiner, and Douglas Purvis, *Economics*, 7th ed. (Harper and Row, 1984).

40. In a recent report it is estimated that the insurance industry as a whole spent $410 million on transaction costs in 1989—primarily in insurance-coverage disputes or in defending the insured—on claims involving hazardous waste sites (both on and off the National Priorities List). The five large PRPs analyzed in the report spent an average of 21 percent of their total outlays (including remediation) on transaction costs, primarily legal fees. Jan Paul Acton and Lloyd S. Dixon, *Superfund and Transaction Costs: The Experiences of Insurers and Very Large Industrial Firms*, R-4132-ICJ (Santa Monica, Calif.: Rand Institute for Civil Justice, 1992), pp. x–xii.

41. For example, see Barnaby J. Feder, "In the Clutches of the Superfund Mess," *New York Times*, June 16, 1991, sec. 3, p. 1; J. Kent Holland, Jr., "Superfund Liability Law Prevents Cleanups," paper presented at the 1990 annual meeting of the American Bar Association.

42. Quoted in the American Insurance Group, *Superfund Issues Forum: Superfund Site Study of Picillo Farm* (Washington, 1991), p. 4.

43. See comments of Senator Alan Simpson, *Congressional Record*, October 3, 1986, p. S4492. See, generally, Kimberly Ann Leue, "Private Party Settlements in the Superfund Amendment and Reauthorization Act of 1986."

44. Following the criminal justice perspective, a corollary to the objective of punishment is the expectation that successful Superfund "prosecutions" will deter irresponsible disposal of toxic wastes in the future.

45. For example, the Environmental Law Institute provided the EPA with a "Draft Enforcement Report" in which it recommended an "effective enforcement" based on a credible threat of detection and punishment: "In order to create a credible threat of detection and punishment with a minimum expenditure of government resource, government must implement punitive actions strategically and successfully over time. If government takes effective punitive action in selected cases and communicates the consequences of its

actions frequently, other members of the obligated community will, over time, fulfill their legal duties 'voluntarily.'" (Environmental Law Institute, "ELI Draft Enforcement Report," sec. 2, p. 4.)

46. Indeed, the compensatory, nonjudgmental nature of the program is often cited by those who defend application of strict, joint and several liability in the face of arguments that such doctrines are unfair. See, generally, Wilkerson and Church, "The Gorilla in the Closet."

47. There have been suggestions that constitutional requirements of due process are violated by CERCLA's *civil* liability structure. See, for example, George C. Freeman, Jr., "Inappropriate and Unconstitutional Retroactive Application of Superfund Liability," *Business Lawyer*, vol. 42 (November 1986), pp. 215–48.

Chapter Three

1. Mutually reinforcing reasons may explain this view: when lawyers are in charge they see legal issues as central, and lawyers may be most sympathetic to approaches that reinforce their control.

2. Mixed funding and *de minimis* buyouts also require (or are thought to require) a substantial investment of administrative effort, because multiple clearances must be obtained and paperwork can be substantial. Given the heavy work load in the region, there is little tolerance for expending time and effort on such activities if the PRPs can be made to pay the whole amount, without *any* concessions by or added administrative burdens on the agency.

3. As Kimberly Leue points out, many of these benchmarks have a statutory origin; the EPA has chosen to use them as measures of organizational performance. Kimberly Leue, "Private Party Settlement in Superfund Amendment and Reauthorization Act of 1986," *Stanford Environmental Law Journal*, vol. 8 (1989), pp. 131–73.

4. In its search for more beans, the region makes regular use of the ability to subdivide sites into "operable units." Although intended to expedite administrative decisionmaking and allow work to commence on discrete parts of a site, this tactic also serves to increase the numbers of reportable activities at a site.

5. For a discussion of similar issues involved in a prosecutorial approach, see Robert Kagan, "Adversarial Legalism and American Government," *Journal of Policy Analysis and Management*, vol. 10 (Summer 1991), pp. 369–406.

6. See, for example, the administrative record in the case of Pristine Belvidere, in which one PRP had to initiate Freedom of Information Act proceedings to get information held by Region V, and the PRPs collectively protested the limited time allotted for comments on the feasibility study. Another Region V example is found in the administrative record in the case of Fields Brook, in which the PRPs could not get the EPA to share environmental studies and went around the agency to get the information from the Ohio EPA.

This situation contrasts sharply with that found in Region III, where sharing information is seen as essential to achieving settlement. See chap. 4.

Our review of administrative records in Superfund cases has uncovered numerous letters complaining about the unwillingness of the region to share information. We were told, for example, to file a Freedom of Information Act form to get a copy of a consent decree. When we asked regional officials about the allegations concerning truncated comment periods, they indicated that they did not think that shortage of comment time was a problem. However, the PRPs in the cases we examined did complain about inadequate time for comment on RI/FSs, RODs, and RD/RAs, and about the reluctance of regional officials to share drafts of documents or other information with them.

7. On the site were a number of buildings and various oil storage facilities. These included two ponds for separating oil, a boiler house, four storage pits, one underground storage tank, thirty-two above-ground oil storage tanks, a retention pond, a freshwater pond, and a greenhouse complex.

8. For purposes of consistency, *EPA* refers throughout this report to the United States Environmental Protection Agency. In discussions of state environmental protection agencies, the acronym *EPA* will precede the state's name, as in Ohio EPA.

9. Indeed, it could be argued that Laskin did reasonably well in the proceedings. One regional official estimated that he ultimately wound up paying less than five thousand dollars in fines. Furthermore, he received from the other PRPs clear title to a new house at another location, because they needed him off the site so work could begin on the cleanup.

10. These included Alvin Laskin and his businesses, and Warren Production Credit. Warren Production Credit had loaned Laskin money for his nursery business, which Laskin used instead to buy waste oil. Warren subsequently sued Laskin for his assets, which included the site. Unfortunately for Warren, it obtained partial title as a result of this legal action and therefore became a PRP.

11. The New Defendants were American Gauge and Machine, Be Kan, Browning Ferris Industries of Ohio, General Motors, Koppers, Rockwell International, and TRW.

12. The original Laskin Task Group later would be augmented by several other large PRPs.

13. The focus of the prosecution strategy on a small group of deep-pocket defendants was made explicit by a Department of Justice attorney in open court: "I would like to point out to the court some of the names of the parties we have represented . . . TRW, General Motors, Rockwell. It is hard to comprehend that the eight parties . . . could not reach a settlement among them for the $2,300,000. . . . Certainly following a settlement of that amount of money the defendants would be welcome to proceed against . . . third parties." Laskin Poplar transcript, U.S. District Court, Cleveland.

14. Laskin Poplar transcript.

15. The Petroleum Exclusion Clauses exempt certain petroleum wastes

from the Superfund program. The actual extent of exclusion was unclear in the statute, however, and gave rise to litigation in a number of Superfund cases. The Petroleum Exclusion Group (as of February 6, 1987) included Ashland Oil, Cleveland Electric, Consolidated Rail, White Consolidated Industries, Shell, Mobil, Sun Refining, Standard Oil, Eliskim, Matlack, and Anchor Motor Freight.

16. A sense of the scope of litigation at this point is illustrated by a motion made in November 1982 by Kaiser Aluminum, asking the court's permission to employ lawyers not admitted to practice in that jurisdiction: "The enormous number of third-party defendants joined in the . . . action, will make it difficult, if not impossible, for each party to obtain a local attorney with the requisite expertise and knowledge to effectively represent clients in this action." The Special Master, noting 660 named parties, waived the local rule.

17. While the government may have hoped that this action would lead to a fast recovery, it was not until three years later, in 1989, that the PRPs settled on the issue of past costs.

18. An exception to this relatively passive role in settlement was a "waste-in list" specifying how much was contributed to the site by the various PRPs, which was developed by Assistant Regional Counsel John McPhee, who, unlike nearly all his colleagues, was with this case for nearly its entire duration. The list, developed from government records and from material obtained in third-party discovery actions, was produced to help the PRPs apportion costs for the purposes of a settlement. This effort was, according to other interviewees, useful in facilitating settlement.

19. The discussion of Ohio's role is drawn from the Ohio EPA administrative record, which was made available to us. This record, unlike the EPA's administrative record, was relatively complete since the Ohio EPA did not have the EPA's policy of removing documents not required to support administrative decisions.

20. The Ohio EPA indicated that Ohio's air toxics policy should apply as an applicable or relevant and appropriate regulation for the cleanup. The Laskin Task Group balked, complaining that it had not been considered in the record of decision, and that the state had not yet published that policy or solicited or considered public comments on it. The Laskin Task Group argued that applying that policy as an ARAR would dramatically extend schedules for completion of the design work and consent decree negotiations.

21. EPA officials contend that the agency now conducts more extensive PRP searches before initiating proceedings.

22. Notification of Hazardous Waste Site document, June 4, 1981. From the administrative record.

23. See site description, Administrative Order by Consent, October 23, 1984, p. 2.

24. The agreement to conduct the RI/FS was formalized in an administrative order by consent issued in October 1984. The government agencies—the EPA and the Michigan Department of Natural Resources—agreed not to sue

the PRPs for matters arising out of the RI/FS, while the PRPs agreed to do the work and submit periodic reports.

25. The Michigan Department of Natural Resources took the position that its water standards should be the applicable cleanup standard. The state agency was adamant that contamination be removed from the groundwater by means of pump-and-treat technology. This position galled the PRPs who pointed out that Michigan, at the time of these demands, had not yet fully developed the administrative rules that defined the requirements of the state superfund act. The PRPs also argued that the issue of groundwater contamination should be considered in light of the location of the site: a swamp from which water was unlikely to migrate.

26. The notification process and its content provide a good example of the reliance on established routines in Region V. When asked why they proceeded as they did, "routine" was the answer given by interviewees. One said, "The overall enforcement policy of the agency is to get the PRPs to do the work, but in this region we always offer the PRPs the opportunity to conduct the RI/FS." Another responded in terms of the same routines: "We have two interfaces with the PRPs. One is at the point where we commence the RI/FS, the other is at the point where we commence to select a remedy. . . . Special notice letters are sent at both stages." In short, regional officials responded that they did what they did in this case because that is what they always do.

27. Figures, drawn from the EPA's CERCLIS data base in 1991, indicate that 22 percent of RI/FSs in Region V are PRP-led, while the figures are 30 percent and 33 percent for Regions III and IV, respectively.

Chapter Four

1. We could not identify a large, complex Superfund case in Region III that had moved to the remedial design/remedial action process in which the final stages of the case were not heavily influenced by Justice Department attorneys who, as will be shown, adopt a forthright prosecutorial stance.

2. The original complaint was amended in 1982 to include injunctive relief under section 106 of CERCLA.

3. This term for a PRP whose liability is ironclad comes from an interview with one of the PRP lawyers involved in the litigation.

4. The congressional survey was conducted by the congressman for whom it was named. Following the controversial events at New York's Love Canal and related hazardous waste sites, manufacturing concerns across the United States were asked to complete a questionnaire in which they indicated where they disposed of various of their toxic chemical wastes. The questionnaires were directed in most cases to plant managers who sometimes merely directed them to the middle managers in their plants who were in charge of waste disposal. The candor of the responses in many cases proved very ex-

pensive for the concerned corporations in subsequent Superfund proceedings.

5. Ward subsequently signed a consent decree in 1983 with the federal government that permitted access to his property by government officials and contractors, gave him protection from lawsuits arising out of actions of the government, and recognized that he would provide certain services in kind, such as provision of trailers for use of personnel on the site and repairs to the fencing. The total value of Ward's contributions was about $37,000. According to his attorney, Ward signed this consent decree in an effort to "build good will with the EPA attorneys."

6. The lead story in the *Philadelphia Inquirer* of September 1, 1983, began, "A single rusting yellow iron bar guards the way. There are no signs to keep out the hunters. No signs warn the teenagers who drink in the broken blockhouse. Nothing stops the toughs who race cars in the soft brown dirt. No one has warned the families who fish and clam nearby. There is only the faint stench and yellow puddles and stark dead trees to warn of the deadly chemicals that have seeped out for more than a decade from the abandoned 50-acre former county landfill, threatening the drinking water for up to 100,000 people."

7. While a health and risk assessment completed by the government's engineering contractor in September 1983 indicated a health risk in only one well (another reportedly had a "significant but safe" level of organic chemicals), community residents and local officials understandably were concerned about the overall safety of their water supply.

8. These included answers, motions to dismiss, briefs on motions to dismiss, cross complaints, amended complaints, and answers to amended complaints.

9. These included interrogatories, requests for production of documents, motions to compel, and briefs relating to those motions. Not included are stipulations for extensions of time and other pro forma filings.

10. ICI sent thirty pages of detailed questions to every manufacturing firm within a one hundred-mile radius of Tybouts. The firms were asked to answer these questions for each manufacturing process and each waste stream at their facility.

11. This material, removed from the river bottom in dredging operations, had the dual advantages of being virtually free and available in large quantity at a location close to Tybouts Corner.

12. While use of dredge spoils in the cap was not specifically mentioned, the EPA technical staff who drafted the ROD told us that they included the diagram only as an example, and believed that use of the dredge spoils would be permitted so long as this alternative material proved to be both impermeable and nontoxic. However, after the ROD was signed and it became reasonably clear that the PRPs were going to build the remedy, control of the case moved from the engineers in the Fund Section of the EPA's Region III to the lawyers in the Enforcement Section. Enforcement lawyers were less amenable

to nontraditional technical solutions than were the engineers in the fund-led side of the office.

13. The government's motion to file an amended complaint was made in late October 1986, and would have added four additional defendants—du Pont, Budd, Witco, and NBF—to the list of primary defendants in the case. These companies defeated this effort by arguing that it was untimely and prejudicial.

14. The second-tier defendants negotiated a deal with the government in which past costs—particularly the costs of building the water supply system for the neighbors of the Tybouts dump and engineering expenses involved in designing the alternate remedy—would not be assessed against them. The implication of this arrangement was that the primary defendants would have to absorb all of these costs, rather than just their negotiated share (79 percent) of them.

15. The major disagreement concerned whether or not the second tier-defendants would pay the insurance and indemnification costs involved in actual construction of the remedy. The consent decree they negotiated with the government had exempted them from these costs.

16. Letter of November 28, 1984, from Judy Dorsey, assistant regional counsel. From administrative record.

17. The problem of the plaintiffs in third-party contribution suits is further complicated by the current confusion over the applicability of joint and several liability to third-party contribution actions in CERCLA cases.

18. Given the presence of government attorneys at each of the scores of depositions taken in the case, together with the need to sift through the hundreds of sets of interrogatories issued and answered, the putative savings of governmental transaction costs may have been illusory.

19. The attorney for the Lands Division of the Justice Department told the federal district judge, "We believe very strongly that we need to build [oversight costs and stipulated penalties] into these decrees against people who are not in the construction business, not trying to do business with the Government, but are unwilling participants in these cleanups, because of the liability Congress has created, and the Environmental Protection Agency has a very, very serious responsibility, your Honor, to the Congress and to the public to make these cleanups happen, to make them happen effectively, and to conserve the resources of the Superfund." Transcript of status conference of March 28, 1988, p. 44.

20. Transcript of status conference of March 28, 1988, pp. 49, 51.

21. The attorney jointly representing ICI and Stauffer complained to the judge: "I'm not privy to the Government's inner dealings, but what you have here is you have got the Department of Justice, which is one entity of the Government, negotiating, then you have got the EPA, which is another entity of the Government, negotiating, and you have their respective intermediate authorities or superiors and their ultimate superiors going up two separate ladders. In addition to that, you have the Super Fund branch, which, as I

understand, is a part of the EPA, but apparently negotiates its own agreements. . . . Maybe the way to consummate this settlement is to get someone
from the Government who's got authority to speak for—there is only one
United States Government, as I understand it—to speak for the United States
Government and to sit down and negotiate this to a conclusion." (Transcript
of status conference of March 28, 1988, pp. 30–31.)

On several occasions in open court, Judge Longobardi suggested that he
would act as a mediator in helping the parties reach settlement: "I'm happy
to step out in the corridor with [the lead Department of Justice attorney] and
whoever wants to be in on this and let's see if we can stitch something together that might work here, if you're willing to proceed in that way." (p. 71).
The Justice Department attorney demurred.

22. A newspaper story describing the scare over the water supply indicated that awareness of the site's use for disposal of paint sludge from nearby
General Motors and Chrysler factories was widespread. See Nancy Kessler,
"One Hundred Families Told of Water Danger," *Wilmington News-Journal*, May
15, 1982.

23. Du Pont and Amchem drums were found at the site. Both companies
successfully argued that the barrels originally had contained end products of
their companies, and thus constituted the waste of the purchasers of those
products.

24. Administrative record.

25. Record of Decision, p. 21.

26. Indeed, several of the participants we interviewed in regard to this
case expressed doubt that this kind of arrangement would be possible under
more recent EPA policies and practices.

27. The GM Request for Preauthorization of CERCLA Funds, pp. 15–16.
Causes of uncertainty listed in this document were: (1) the number of drums
on the surface prevented the use of magnetometer surveys to determine the
number of buried drums; (2) the cost of construction and operation of the
"pump-and-treat" system depended on a number of factors that could not be
known until additional studies and engineering design were further along;
(3) the most effective configuration of the soil-flushing system could not be
set out on the basis of the then-current level of information, except in the
broadest terms.

28. The EPA suggested that less, not more, specificity was necessary, that
it was "not practical to design the groundwater recovery or treatment systems
until we establish the *need for*, and the *extent of*, offsite groundwater recovery
pumping." Agency letter of April 2, 1986, to Joseph Chu, from administrative
record. Emphasis in original.

29. Between the time when the Harvey and Knott ROD was signed and
the final settlement discussions were taking place, Congress had passed the
Superfund Amendments and Reauthorization Act of 1986 (SARA). The EPA
was not willing to "grandfather" the Harvey and Knott ROD into CERCLA
and thus evade the newer, more stringent standards of SARA concerning the
critical how-clean-is-clean issue. The issue ultimately was resolved by means

of a clause in SARA which allows alternative concentration limits to be applied in place of the ARARs required by SARA.

Chapter Five

1. For discussions of Region IV in the policy literature, see Ann O'M. Bowman, "Intergovernmental and Intersectoral Tensions in Environmental Policy Implementation: The Case of Hazardous Waste," *Policy Studies Review*, vol. 4 (November 1984), pp. 230–44. See also Bowman, "Hazardous Waste Cleanup and Superfund Implementation in the Southeast," *Policy Studies Journal*, vol. 14 (September 1985), pp. 100–10; and James E. Lester and others, "Hazardous Wastes, Politics, and Public Policy: A Comparative State Analysis," *Western Political Quarterly*, vol. 36 (June 1983), pp. 257–85.

2. Thus, even the most prosecutorial among them reportedly begins discussions with PRPs on the assumption that "you're not here because of anything that you intentionally did wrong, but because the law says you have a responsibility to pay for cleanup of your waste."

3. Removals can be financed by PRPs, as part of a consent decree of administrative order on consent. However, the negotiations necessary to achieve this form of financing necessarily slow the cleanup process. Region IV's approach typically has emphasized use of federal funds for the removals, to be followed in some cases by efforts to obtain reimbursement from PRPs.

4. According to one interviewee, at a formative stage in the history of Region IV, the idea was "to clean up the environment and spend money"; the key individuals in charge of the region were more concerned with cleaning things up than with legal niceties.

5. With the exception of several major sites in Kentucky, most of the sites in the region, according to one interviewee, "are smaller than, say, [those in] Region V. . . . We have a lot of cases where there are no generators but landowners, which involve more complicated theories of liability." A number of those landowners are small businesses or penurious municipal PRPs, and thus are not viable PRPs. Conversely, some sites are difficult because of the absence of an owner-operator, even though some small generators may be present.

6. Region IV is the site of the Chemical Waste Management Corporation operation, the largest Resource Conservation and Recovery Act disposal facility in the nation, as well as the GSX RCRA facility in South Carolina.

7. There are, however, headquarters approvals to be obtained. We understand that these authorizations frequently involve pitched battles, depending on the national implementation policy currently being espoused.

8. Comprehensive Environmental Recovery, Compensation, and Liability Act of 1980, P.L. 95-510.

9. See EPA Inspector General Report, Region IV, "Review of Region IV's Management of Significant Removal Actions," September 28, 1988.

10. Moreover, in states such as Georgia which have historically resisted

the listing of sites on the NPL, the removal sometimes has provided the impetus to clean up and thus avoid political conflict with the state.

11. Apparently, the estimated cost of the cleanup was below the $2-million cap on emergency removals, despite the fact that the cost estimates for the remedy chosen in the ROD (and implemented in the removal) ranged from $2.2 million to $3.1 million.

12. The decision to relocate some families as a result of the 1978 gas problem caused others to demand that their properties also be purchased and their families relocated. The residents contested various conclusions in the RI/FS, including the assertion that the dump was only forty to fifty feet deep. Interestingly, the EPA did not do test borings in the landfill because of "potential health hazards" that might result.

13. The most thorough options, which would have involved extensive incineration, or the removal of contaminated soils, would have cost $418.1 million and $649.2 million respectively.

14. Although this remedy came at a time when the relationship between Kentucky environmental officials and the EPA was in decline, the state did not contest the remedy.

15. The region did not maintain an administrative record on removals at this time, and we do not have reliable information on how these irregularities were dealt with.

16. As another attorney saw it, negotiations over cost recovery are quite different from negotiations between PRPs over taking the lead on a remedy. In the latter situation, "All you have is the estimates in the ROD, the estimates of your consultants, and a wing and a prayer," this attorney said.

17. The EPA carried out the O&M activities for the first two years after the remedy. An additional twenty-eight years were required by the ROD. The Louisville and Jefferson County Metropolitan Sewer District (MSD) thought it could do the work for less than the EPA estimates and offered to do the O&M as its share of the deal. The consent decree was written to permit the rest of the negotiation to move forward by providing that the PRPs would divide up the half-million dollar estimated cost for O&M activities if the MSD failed ultimately to agree on performance of the work.

18. This concern was raised by two PRP counsel and noted specifically by Ford in its Memorandum in Opposition to Government's Motion to Enter Consent Decree, February 7, 1991, p. 14. Neither the EPA nor the Department of Justice confirmed Ford's suspicion, however. They asserted that "enforcement considerations" required delays and that it was not unusual for a substantial period to elapse between conclusion of the main negotiations in a case and the lodging of the decree. The EPA and the Justice Department argued that the decree could not possibly have been lodged before mid-1989. *United States v. Ben Hardy*, Reply of the U.S. Government to Objections to Entry of the Consent Decree, p. 19, note 9. Government respondents rejected that contention.

19. Ford had been considered as a PRP earlier in the case but was dropped

from the proceedings when it appeared that its wastes consisted only of construction debris and nonhazardous materials. Even so, Ford offered in late 1987 and early 1988 to discuss a *de minimis* settlement. (Mark D. Edie to Anna C. Thode, January 2, 1991, pp. 1–2.) Based on information from a truck driver, the EPA later concluded that Ford had deposited paint sludge at Lee's Lane, and reinstated the company as a PRP.

20. Mark D. Edie to Anna C. Thode, January 2, 1991, p. 3. The companies also argued that issuance of the consent decree had been delayed at the government's instigation and that there was no "time value of money" issue involved because none of the other settling parties had yet made any reimbursement to the government.

21. The 4 percent figure was the amount to be paid by Waste Management of Kentucky, successor to the firm that transported Ford and Dow Corning wastes to Lee's Lane. *United States* v. *Ben Hardy*, Ford's Memorandum in Opposition to Government's Motion to Enter Consent Decree, p. 22, note 10.

22. The firm's officials kept up the pretext that waste was being incinerated, even going so far as to substitute tap water for the waste in a test burn conducted at the insistence of the Georgia Environmental Protection Department, some weeks after the Anniston waste had been flushed into municipal sewers. *United States* v. *Everett O. Harwell and Eugene Roy Baggett*, Sentencing Memorandum, p. 5.

23. Ironically, the firm met with Georgia authorities and agreed to pay a $50,000 penalty for violating air emission standards, even though the waste in question had been flushed it into city sewers.

24. Later, the company tried other expedients. Waste drums were placed in tractor-trailer trucks and moved to various locations off site. An anonymous caller told the Georgia EPA that the trucks were being used to evade inspection. The state found twenty-two trucks filled with drums.

25. Criminal action against Harwell and Baggett eventually was brought in December 1985. The two pleaded guilty to two felony counts in October 1986. Harwell was sentenced to three years' imprisonment and a $20,000 fine. Baggett was given eighteen months in jail and a $10,000 fine. *United States* v. *Everett O. Harwell and Eugene Roy Baggett*, Doc. No. CR85-28 R.

26. The Davis brothers had assisted the prosecutors with the understanding that they would not be targeted by other agencies. Hence, from the beginning, the technical owner-operator of the site was virtually immune from EPA action.

27. Region IV contended that passage of the Superfund Amendments and Reauthorization Act in 1986 gave them until late 1989 to file the cost recovery suit. The primary evidence consisted of records showing that firms had sent certain kinds of wastes to SWT, manifests from SWT, and data indicating that the same type of waste was found at Davis Farm. Unfortunately, the manifests did not make clear how much of the waste received by SWT was deposited at Davis Farm. Both Justice Department and PRP attorneys expressed doubts about the reliability of an apportionment based on this data.

28. The federal facilities included the Tennessee Valley Authority (TVA), Letterkenny Army Depot, Anniston Army Depot, the Naval Air Engineer Center, and the Pittsburgh Energy Technology Center.

29. One of the private-sector PRP attorneys designed a microcomputer spreadsheed for all PRPs, based on the EPA data. The PRPs then supplemented this data to make share allocations more accurate. These spreadsheet data were used in reaching both federal and private PRP settlements.

30. CERCLIS data through 1990 indicate only three NBARs have been produced for all NPL sites in the United States. Some interviewees suggested that PRPs are reluctant to request NBARs because they fear they may ultimately be bound by them (despite their supposedly nonbinding character), and because they assume that the EPA will generate a 100 percent allocation in situations where some form of mixed funding is more appropriate.

31. When the information is sufficiently detailed, it may also reduce the need for multiple negotiations to resolve questions concerning issues such as *de minimis* settlements. If multiple steering committees are needed, the volumetric-share data make it easier to coordinate settlement totals.

Chapter Six

1. For a discussion of the creation of the Superfund program, see Marc K. Landy, Marc J. Roberts, and Stephen R. Thomas, *The Environmental Protection Agency: Asking the Wrong Questions* (New York: Oxford University Press, 1990). A critical review of overall hazardous waste policy is found in Daniel Mazmanian and David Morell, *Beyond Superfailure: America's Toxics Policy for the 1990s* (Westview Press, 1992).

2. For example, EPA figures show that the number and value of Superfund settlements rose from 114 ($175.2 million) in FY 1987 to 218 ($1,033.7 million) in FY 1989. U.S. EPA Office of Enforcement and Compliance Monitoring figures from EPA, *Progress toward Implementing Superfund, Fiscal Year 1989* (Washington, June 20, 1991).

3. Martin Rein and Francine Rabinovitz, "Implementation: A Theoretical Perspective," in W. D. Burnham and M. W. Weinberg, eds., *American Politics and Public Policy* (MIT Press, 1978).

4. David Weimer and Aidan Vining, *Policy Analysis: Concepts and Practice* (Prentice Hall, 1989), p. 1.

5. See, for example, Environmental Defense Fund and others, *Right Train, Wrong Track: Failed Leadership in the Superfund Cleanup Program* (Washington, 1988); Frank Lautenberg and Dave Durenberger, *Lautenberg-Durenberger Report on Superfund Implementation: Cleaning Up the Nation's Cleanup Program* (Washington: Senate Subcommittee on Superfund, Ocean, and Water Protection, 1989).

6. In some studies, records of decision were examined to determine the extent of reliance on "cap and fence" alternatives versus permanent—and more expensive—options. The underlying assumption was that the latter are

always preferable to the former. See, for example, Environmental Defense Fund and others, *Right Train, Wrong Track*; Office of Technology Assessment, *Are We Cleaning Up?: Ten Superfund Case Studies* (Washington, 1988).

7. We briefly compare the implementation strategies in terms of cost of the remedies selected at the end of this section.

8. Clean Sites, *Improving Remedy Selection: An Explicit and Interactive Process for the Superfund Program* (Alexandria, Va., 1990), preface. We indicated earlier that Clean Sites provided the grant for producing the case studies reported in this work. Clean Sites did not, however, constrain our perspective on remedy selection or on any other substantial matter. The cited report was chosen because it is one of the few pieces to articulate criteria for assessing remedy selection, and because its recommendations have been found convincing by a wide range of participants.

9. In Tybouts Corner, however, informal understandings between the PRPs and the program staff in the agency concerning the use of dredge spoils in the Resource Conservation and Recovery Act (RCRA) cap came unhinged when control of the case moved to the lawyers in the regional counsel's office and the Justice Department, after the ROD was signed. The accommodation style of decisionmaking in regard to technical issues changed to a more prosecutorial mode. In neither case was there much community or state involvement.

10. One contrary suggestion in the critical literature on Superfund remedy selection is that remedies that arise from negotiations with PRPs are less protective of human health and the environment than are remedies funded by the federal government. There is an expectation in the PRP community that fund-led cleanups are more *expensive* than PRP-led cleanups, although the higher cost of EPA cleanups is sometimes ascribed to government inefficiency rather than to any superiority of the actual result. See Environmental Defense Fund and others, *Right Train, Wrong Track*; and Lautenberg and Durenberger, *Lautenberg-Durenberger Report*.

11. Such a result would be consistent with studies of the relationship of the perception of fairness to compliance in other legal contexts: in several settings it has been shown that litigants who feel they have been dealt with fairly and who have had a chance to have their say are more likely to comply with adverse decisions than those who feel that procedures in their cases prevented them from participating meaningfully. See, for example, Craig A. McEwen and Richard J. Maiman, "Mediation in Small Claims Court: Achieving Compliance through Consent," *Law and Society Review*, vol. 18, no. 1 (1984), pp 11–50.

12. However, public participation was insubstantial in virtually all of our case studies, regardless of implementation strategy.

13. Clean Sites, *Improving Remedy Selection*, p. iv.

14. Once the consent decree is signed at a site where the cleanup is to be conducted by PRPs, these parties are formally responsible for an acceptable remedy, and must demonstrate to the EPA that all ARARs have been met.

15. Our two Region III cases did not involve ARARs, because both cases

took place between promulgation of the CERCLA requirements in 1980 and enactment of the more stringent SARA provisions in 1986. The order in which decisions were taken, however, was similar to that in Region V: the RI/FS and ROD processes preceded decisions on the actual cleanup standards that would be applied. In Harvey and Knott, General Motors and the EPA agreed on a process by which standards would be generated as the cleanup progressed.

16. Clean Sites, *Improving Remedy Selection*, p. 31.

17. This potential was recognized by Clean Sites. See *Improving Remedy Selection*, preface.

18. See, for example, Environmental Defense Fund and others, *Right Train, Wrong Track*; and Office of Technology Assessment, *Are We Cleaning Up?*

19. In the Lee's Lane Landfill case, regional decisionmakers faced with several ROD alternatives coincidentally chose the one that permitted them to do the work immediately, and for no more than the allotted amount. Again, it appears that available government resources may have pushed selection of that option. The fear is, of course, that restrictions on the use of fund monies in removal may push decisionmakers to sacrifice thoroughness for speed.

Interestingly, government decisionmakers were the ones who pressed for relatively inexpensive remedies in these public works cases. Our interviews with the Superfund community have uncovered one consistent relationship: those who pay the costs usually press for cheaper solutions. So, when states must pay part of cleanup costs, they press for less expensive remedies than when the PRPs are expected to pay the full costs. The behavior of PRPs is also consistent in this regard when they have to pay. We were told that at times when the fund was running low, even federal EPA preferences at fund-led sites were affected by the prospect of exhausted resources.

20. Later in this chapter we assess separately the issue of transaction costs.

21. See Kagan and Scholz, "The 'Criminality of the Corporation' and Regulatory Enforcement Strategies," in Keith Hawkins and John Thomas, eds., *Enforcing Regulation* (Kluwer/Nijoff, 1984).

22. When a set of conditions is present—that is, the removal constitutes the bulk of the remedial action, costs can be allocated among a substantial number of PRPs and individual shares remain modest, the data on allocation is credible—cost recovery is likely to work well.

23. See EPA figures reported in the chart, "Breakdown of Superfund Activity," *National Law Journal*, February 18, 1991, p. 39. See also Alfred R. Light, "Superfund: Evaluations and Proposals for Reform," paper presented at the 1990 annual conference of the American Society for Public Administration, p. 2.

24. The EPA has had difficulty meeting these objectives. See Lautenberg and Durenberger, *Lautenberg-Durenberger Report*, pp. 84–86. There are indications that decisionmaking since SARA has actually been slower than under CERCLA. The Center for Hazardous Waste Management found that both

RODs and RI/FSs proceeded much more slowly in 1987 and 1988 than they had before passage of SARA in 1986. See Light, "Superfund: Evaluations and Proposals for Reform."

25. See Light, "Superfund."

26. According to one report, post-SARA RI/FSs took an average of thirty-three months to complete. See Lautenberg and Durenberger, *Lautenberg-Durenberger Report*. See also Light, "Superfund."

27. The contractors who conducted the RI/FS were criticized for their initial slow pace at Tybouts; the PRPs were not substantially involved at that point. After the ROD was signed, control of the case moved to Justice Department attorneys. Proceedings subsequently were characterized more by a prosecution than an accommodation approach.

28. See Clean Sites, *Improving Remedy Selection*.

29. See Barnaby J. Feder, "In the Clutches of the Superfund Mess," *New York Times*, June 16, 1991, sec. 3, p. 1.

30. For a discussion of transaction costs, see Jan Paul Acton and Lloyd S. Dixon, *Superfund Transaction Costs: The Experiences of Insurers and Very Large Industrial Firms*, R-4132-ICJ (Santa Monica, Calif.: Rand Institute for Civil Justice, 1992). The Rand study stands as the only one which, even for a limited set of firms, has managed to include an estimate of private-party transaction costs. We have transaction cost estimates for the government—although not for the PRPs or their insurers—for four of the cases we examined. Total enforcement costs as of February 1992, according to informal EPA estimates provided to us, were: Laskin Poplar, $950,000; Cliffs Dow, $271,000; Harvey and Knott, $482,000 (this figure includes the legal costs involved in proceeding against the major nonsettler, Chrysler); Tybouts Corner, $905,000. We have no estimates for the Region IV cases.

31. In addition, the settling PRPs may—since a 1991 court decision—be able to recover their legal fees in private-party cost recovery actions against nonsettlers. See United States Court of Appeals for the Eighth Circuit, *General Electric* v. *Litton Industrial Automation Systems, Inc. and Litton Industries, Inc.*, No. 89-2845, 1991. We were told that the Department of Justice apparently worried that this capacity would encourage PRPs to litigate in the expectation that some of their costs after settlement could be later recovered.

32. As suggested in chap. 4, cooperation characterized the entire case, with extensive PRP participation in the RI/FS, and mixed funding. The case required no recourse to a judge for resolution of disputes (although the consent decree provided for such recourse in the case of irreconcilable differences).

33. There may be a general reluctance to devise overall strategies for a site on a systematic basis. For example, the General Accounting Office found that "EPA . . . often does not develop or document a minimum settlement target at any point in the negotiation process. . . . Regional enforcement staff told us that they did not develop written bottom-line strategies because EPA policy does not require them to do so." GAO, *Superfund: More Settlement Authority*

and EPA Controls Could Increase Cost Recovery, GAO/RCED-91-144 (Washington, July 1991), p. 26.

34. Ibid., p. 26.

35. For example, large industrial waste dumps with multiple PRPs are more likely to be found in the Northeast and Midwest than in the Far West or the Southeast; government facilities are more common in the West; small size and single-party responsibility characterize many of the sites in the Southeast.

36. Kagan and Scholz, "The 'Criminology of the Corporation' and Regulatory Enforcement Strategies," pp. 85–86.

Chapter Seven

1. This is David Weimer's apt use of an old saw for such direct policy advice.

2. See Jeffrey Pressman and Aaron Wildavsky, *Implementation* (University of California Press, 1973).

3. See Eugene Bardach, *The Implementation Game: What Happens after a Bill Becomes Law* (MIT Press, 1977), for his discussion of specific games. Among the most relevant are "not my problem," "easy money," and "tenacity."

4. For a discussion of the "agency loss" problem, see Terry Moe, "The New Economics of Organization," *American Journal of Political Science,* vol. 28 (November 1984), pp. 739–77.

5. Elsewhere we explore the relationship between the discretion that Congress grants the EPA and the agency's uses of it. See Robert Nakamura, Thomas Church, and Phillip Cooper, "A Blip on the Radar Screen: Implementation of the Medical Waste Tracking Act of 1988," *Journal of Health Politics, Policy and Law,* vol. 17 (Spring 1992), p. 299.

6. The term *advocacy framework* comes from schools of social work where different professionals are taught to "advocate" for specific aspects of a client's interest. The idea of applying this term in this context is our own.

7. May and Williams deal with what they term "shared governance," policies that require participation by multiple government units and the consequent increased potential for disagreements over ends and means. Peter J. May and Walter Williams, *Disaster Policy Implementation: Managing Programs under Shared Governance* (Plenum Press, 1986).

8. The phrase is Charles Lindblom's, from his "Science of Muddling Through," *Public Administration Review,* vol. 19 (Spring 1959), pp. 79–88.

9. Bardach, *The Implementation Game;* and Martin A. Levin and Barbara Ferman, *The Political Hand: Policy Implementation and Youth Employment Programs* (Pergamon Press, 1985).

10. The term *street level bureaucracy* is Michael Lipsky's. See *Street Level Bureaucracy: Dilemmas of the Individual in Public Service* (New York: Russell Sage Foundation, 1980).

11. The term is from Paul Sabatier, "Top-Down and Bottom-Up Ap-

proaches to Implementation Research," paper presented to the 1985 world conference of the International Political Science Association.

12. See Bardach, *The Implementation Game*, and Levin and Ferman, *The Political Hand*, for a discussion of the fixer role.

13. For a discussion of an unsuccessful attempt to use ADR techniques to facilitate a hazardous waste cleanup in New York State, see Robert T. Nakamura, Thomas W. Church, Jr., and Phillip J. Cooper, "Environmental Dispute Resolution and Hazardous Waste Cleanups: A Cautionary Tale of Policy Implementation," *Journal of Policy Analysis and Management*, vol. 10 (Spring 1991), pp. 204–21.

14. See Christopher Leman, "The Forgotten Fundamental: Successes and Excesses of Direct Government," in Lester Salamon, ed., *Beyond Privatization: The Tools of Government Action* (Washington: Urban Institute Press, 1989).

15. We learned of this effort in Region III interviews.

16. Such a suggestion was made by Clean Sites. See Clean Sites, *Improving Remedy Selection: An Explicit and Interactive Process for the Superfund Program* (Alexandria, Va., 1990).

17. A recent example of this conflict had Justice Department attorneys intervening in a private-party cost recovery action—following a settlement by the PRPs with the EPA—to prevent private-party recovery from the nonsettlers, on the theory that the government had first claim on their contributions to the settlement (when, of course, it got around to seeking them).

18. See Thomas W. Church and Milton Heumann, "The Underexamined Assumptions of the Invisible Hand: Monetary Incentives as Policy Instruments," *Journal of Policy Analysis and Management*, vol. 8 (Fall 1989), pp. 641–57.

19. This point is made in Clean Sites, *Making Superfund Work: Recommendations to Improve Program Implementation* (Alexandria, Va., 1989) p. 34: "Measures of 'success' for the Superfund program must address actual program accomplishments ('outputs') rather than intermediate steps on the way to cleanup ('inputs') as is currently the case. A focus on inputs rewards effort and intermediate steps that lead towards the ultimate objective; a focus on output rewards accomplishment, achievement of the ultimate goal—site remediation. Success obviously depends on effort, so that it is rational to reward effort. The risk in rewarding effort, however, is that the incentives lead to achieving intermediary milestones in unanticipated or undesired ways that do not actually result in the achievement of the organization's true goals."

20. This lack of information handicaps attempts to assess the aggregate costs of the Superfund program. See Milton Russell, E. William Colglazier, and Mary R. English, *Hazardous Waste Remediation: The Task Ahead* (University of Tennessee, Waste Management Research and Education Institute, 1991).

21. Katherine N. Probst and Paul R. Portney, *Assigning Liability for Superfund Cleanups: An Analysis of Policy Options* (Washington: Resources for the Future, June 1992), p. ix. See also Jan Paul Acton and Lloyd Dixon, *Superfund Transaction Costs* (Santa Monica, Calif.: Rand Institute for Civil Justice, 1992).

22. The term is Marc Galanter's. See "Why the 'Haves' Come Out Ahead: Speculations on the Limits of Legal Change," *Law and Society Review*, vol. 9 (Fall 1974), pp. 95–160.

23. See Thomas Schelling, *A Strategy of Conflict* (Harvard University Press, 1960), for his discussion of continuing and intersecting negotiations.

24. Some of this caution probably results from the painstaking process of defining newly defined legal powers through extensive litigation. Superfund's liability doctrine, for example, underwent years of testing and definition in the courts. As new issues emerge, testing continues. Thus, while government starts with a substantial store of legal rights conferred by statute, effectuating those rights usually requires a considerable investment by administrative agencies. Once those legal tools are established, government is encouraged to guard them jealously, often at the expense of the flexibility required for settlement.

25. We saw, for example, that some prosecution-oriented lawyers in Region V believed that deviations from a full application of joint and several liability in Superfund were somehow immoral. Similarly, we saw how rigid application of the liability doctrine can cause PRP intransigence in situations when they argue that they are not at fault.

26. Landy and others fault Superfund for making cleanup standards a participatory issue involving local publics only to the extent that they tell decisionmakers what they want done without the corresponding need to consider the costs. Laurence Lynn criticizes the view of participation that underlies this position. Laurence E. Lynn, book review of Gerald Jacob, *Site Unseen: The Politics of Siting a Nuclear Waste Repository* (University of Pittsburgh Press, 1990), Howard Kunreuther and Rajeev Gowda, eds., *Integrating Insurance and Risk Management for Hazardous Wastes* (Kluwer Academic Publishers, 1990), and Marc Landy, Marc J. Roberts, and Stephen R. Thomas, *The Environmental Protection Agency: Asking the Wrong Questions* (New York: Oxford University Press, 1990). See Lynn's review in the *Journal of Policy Analysis and Management*, vol. 11, no. 1 (1992), pp. 133–37. According to Lynn, "It is nothing less than utopian, this notion. . . . that if members of policy elites could explain the relevance of their benefit-cost analyses and risk assessments to 'the public,' if 'the public' could be given a coherent account of the trade-offs they face, if self-interested actors and establishments who remain intransigent could be disempowered in favor of an inquiring and open-minded citizenry, public policies more clearly reflective of the 'public interest' would result." Lynn quotes Landy and coauthors: "Costs [were] routinely . . . discussed as if someone else could be made to pay for them" (p. 279). Landy and his coauthors fault "incentive systems and organizational structures that operated within the agency and within the executive branch as a whole." (pp. 283, 284).

27. A simple theoretical exercise illustrates this point. Suppose the cash value of a Superfund settlement would not have to be spent on actual cleanup. Instead, the $30-million cost of the average remediation could be spent on reducing any environmental risk. Would the EPA, Congress, or the local community spend that sum on cleaning up a site? Or would they instead

spend it on reducing air pollution or radon exposure, or on some other environmental activity? In many instances, we suspect that loosening the restrictions on the use of that money would produce different spending patterns.

28. Another important but far less controversial aspect is the strict liability component—the assignment of liability for cleanup expenses without the necessity of finding negligence or fault on the part of a PRP.

29. 572 F. Supp. 802 (S.D. Ohio, 1983). For a full description of the history of joint and several liability, see W. Freeman, *Joint and Several Liability* (Stoneham, Mass.: Butterworth Legal Publishers, 1987).

30. See, for example, Note, "Generator Liability under Superfund for Cleanup of Abandoned Hazardous Waste Dumpsites," *University of Pennsylvania Law Review,* vol. 130 (May 1982), pp. 1229–80; Note, "Joint and Several Liability for Hazardous Waste Releases under Superfund," *Virginia Law Review,* vol. 68 (May 1982), pp. 1157–95.

31. There are grounds, however, to doubt the value added by JSL in a scheme in which strict liability—liability without fault or negligence—already is applicable. See William Wilkerson and Thomas Church, "The Gorilla in the Closet: Joint and Several Liability and the Cleanup of Toxic Waste Sites," *Law and Policy,* vol. 11 (October 1989), pp. 425–29; Steven Shavell, *Economic Analysis of Accident Law* (Harvard University Press, 1987).

32. House Judiciary Committee, *Superfund Reauthorization: Judicial and Legal Issues* (GPO, 1985).

33. For an instructive discussion of potential alternative liability frameworks for Superfund, see Katherine N. Probst and Paul R. Portney, *Assigning Liability for Superfund Cleanups: An Analysis of Policy Options* (Washington: Resources for the Future, 1992).

34. A fair-share standard was proposed by Senator Al Gore, Jr., and was applied in a few early Superfund cases, in particular *U.S.* v. *Stringfellow,* 14 ELR 20, 385 (C.D. California, 1984). For discussion, see Note, "Generator Liability under Superfund for Cleanup of Abandoned Hazardous Waste Dumpsites," *University of Pennsylvania Law Review,* vol. 130 (May 1982), pp. 1229–80; Douglas F. Brennan, "Joint and Several Liability for Generators under Superfund: A Federal Formula for Cost Recovery," *UCLA Journal of Environmental Law,* vol. 5, no. 2 (1986), pp. 101–35. However, the joint and several standard was ultimately upheld in the courts.

35. Several European countries conduct successful cleanup programs without joint and several liability. See Andrew Lohof, "The Cleanup of Inactive Hazardous Waste Sites in Selected Industrialized Countries," Discussion Paper 069 (Washington: American Petroleum Institute, 1991); Raymond Kopp, Paul Portney, and Diane DeWitt, "International Comparisons of Environmental Regulation," Discussion Paper QE90-22-REV (Washington: Resources for the Future, 1990).

36. See Lohof, "The Cleanup of Inactive Hazardous Waste Sites in Selected Industrialized Countries"; Kopp and others, "International Comparisons of Environmental Regulation."

37. See Robert Nakamura, Thomas Church, and Phillip Cooper, "A Blip

on the Radar Screen: Implementation of the Medical Waste Tracking Act of 1988," *Journal of Health Politics, Policy, and Law,* vol. 17 (Summer 1990), pp. 299–329.

38. See Eugene Bardach and Robert Kagan, *Going by the Book: The Problem of Regulatory Unreasonableness* (Temple University Press, 1982).

Index